The Solar Generation

The Solar Generation

Childhood and Adolescence of Terrestrial Photovoltaics

Philip R. Wolfe

IEEE PRESS

WILEY

Published by John Wiley & Sons, Inc., Hoboken, New Jersey.
Published simultaneously in Canada.

For general information on our other products and services or for technical support, please contact our Customer Care Department within the United States at (800) 762-2974, outside the United States at (317) 572-3993 or fax (317) 572-4002.

Wiley also publishes its books in a variety of electronic formats. Some content that appears in print may not be available in electronic formats. For more information about Wiley products, visit our web site at www.wiley.com.

Library of Congress Cataloging-in-Publication Data is available.

ISBN: 978-1-119-42558-8

Printed in the United States of America.

V080722_042718

This book is dedicated to all those
– described herein or otherwise –
who struggled against the status quo
to establish the terrestrial photovoltaics sector;
and build the foundations of
a major clean energy technology.

In memory of friends
Bill Yerkes,
Joseph Lindmayer,
Brian Harper,
John Bonda,
and those others who didn't live
to see the extent of their success[1]

---

1 see Section 10.3.

Contents

A detailed Table of Contents is provided in Appendix E.

Foreword

Behold the blessed vision of the sun, no longer pouring his energies unrequited into space, but by means of photoelectric cells and thermopiles, these powers gathered into electric storehouses, to the total extinction of steam engines and the utter repression of smoke.

Rollo Appleyard, 1891 [1]

It is hard to imagine, now that solar power is an indispensable part of the global energy mix, that it was for so long discounted as an irrelevance. The accepted wisdom for several decades was that solar generation is an expensive, unreliable, small-scale niche application.

True, some of these criticisms had weight in the early years. At first a little imagination or faith was needed to see the true promise of a sustainable clean energy future. Fortunately, the early solar industry attracted pioneers with the vision to see beyond any temporary shortcomings. This book traces their journey as they laid the foundations for the significant business that solar energy has now become.

Solar cells had already been used to power spacecraft from the late 1950s, but widespread use on earth was an entirely different proposition. The terrestrial industry started with technology that was 1000 times too expensive, and production capacity maybe one billionth of what would be needed. This is the story of how far we got in the first quarter century.

It was not attained through the superhuman effort of one individual, but by collective achievement, even though it might not have felt at the time that everyone was working together. There was no single "father of solar power," but rather dozens of inspirational individuals, and a lot of good companies, as you will discover.

You will see how many of the same names pop up in different places as companies, technologies, and applications rise, fall, and transform during the evolution of the sector. It is remarkable how the majority of these people, having migrated into the industry from all walks of life, became captivated by the wonder of solar, and stayed with it for the rest of their working lives.

Who Needs This Book?

Solar energy today is a significant, but secondary, global contributor. I confidently predict that it will be the world's largest energy source within my lifetime.[1]

Curious people, inside the sector and beyond, will want to know how it all began, and this book sets out to tell that story, from the start of solar power usage on earth to the time when widespread deployment was just beginning.

Why Me?

I had the good fortune to fall accidentally into the solar energy sector back in 1975, when it was just getting started. The companies and organizations, in which I was involved,[2] were associated with many of the leading solar businesses and researchers, particularly in Europe and North America.

However, as will become apparent, it is others who have made great achievements on which my own slight contribution has depended. I am lucky enough to have known many of the real movers and shakers and dealt with the leading companies.

Hopefully, I have managed to be objective, while preserving the benefit of having been there at the time. At the risk of a bad attack of schizophrenia, I refer to myself in the first person, as the author of this book; and to Philip Wolfe, in the third person, as a participant in the early PV era.

Volumes, Sections and Chapters, References, and Jargon

This book will be of interest to a wide range of readers and I have tried to structure it in the way that would be of greatest use:

Part I, the "biography" of terrestrial photovoltaics, tracks the evolution of research, technology, markets, prices, the industry, and so on.

Part II is called the "encyclopedia" with a Who's Who of the most influential people and profiles of key companies, events, and developments.

---

1 If I share my parents' longevity, this may extend toward 2050.
2 Mentioned in the profile in Section 10.2.

Part III is the reference section with glossaries, bibliography, and a really comprehensive index. You can find everything in there.

Each part is in chapters (listed in the contents summary above) and each chapter has a number of sections. A more detailed contents list is included in Chapter E of Part III.

I find that a forest of footnotes and references at the bottom of every page makes books harder to read. With a few exceptions, I have put such notes and references at the end. The reference numbers within brackets in the narrative relate to the reference list in Chapter B of Part III. A few pages have footnotes in numerals in superscript. Superscripted[qv] after a name denotes people, organizations, and events individually profiled in Part II.

This is not a scientific textbook, and I have aimed to avoid using lots of acronyms, jargon, and formulas. If, however, you find yourself baffled about what an "MPPT" does, when "grid parity" is reached, or what the "Shockley Queisser Limit" limits, just go to the Glossary in Part III.

Name Checks and Acknowledgments

It has been such a pleasure writing this book and revisiting so many former colleagues around the world (see Chapter A in Part III). It has been more difficult trimming the list of those to be profiled to a sensible length. The intention was to select those who have been most influential, not necessarily those who were most visible. I have undoubtedly committed errors and omissions out of sheer ignorance. For all such cases, let us hope that enough people buy this book to justify further editions, when these mistakes can be rectified.

Hopefully, these pioneers will forgive me for naming them mostly without their professorships, doctorates, and so on. Where these do appear, they are at the level most prevalent during the timescale of this book – it may say Dr. here, but they are probably at least a Prof. by now! Many of the companies mentioned have longer names or adopt special typestyles; I ask their marketing departments, too, to forgive me for sticking to the parlance most commonly used at the time. Individuals profiled in Chapter 10 and organizations in Chapter 11 are annotated with "qv" when they appear in other parts of the book.

Many, many people gave me access to centuries of accumulated knowledge about photovoltaics during the writing of this book. It needs a whole chapter (Chapter A in Part III) to acknowledge them all. I must give a special mention to Bernard McNelis[qv] for allowing me free rein in his extensive archive and photo library. Thanks also to Martin Green[qv] and Dave Carlson[qv] for checking the technological details, to Sabrina Wolfe for proofreading it all and helping me make it a much tidier and more literate work, and to Mary Hatcher, Editor, and

Vishnu Narayanan, Senior Project Editor. Finally, thanks to Alison Wolfe for sitting through many long PV reminiscences, and generally living with this book for all these months.

I have done my best to incorporate their suggestions, and to get the facts right, but the inevitable mistakes are all mine, not theirs. Readers finding substantive errors and omissions are encouraged to notify me through the book's website [234], so that these can be rectified in any future editions.

Part I

A Biography of Terrestrial Photovoltaics

"The silicon solar cell may mark the beginning of a new era leading eventually to the realization of one of the mankind's most cherished dreams – the harnessing of almost limitless energy of the sun for the uses of the civilization."

New York Times, 1954 [2]

Solar power has now become an indispensable part of the global energy mix, so it is easy to forget that less than a generation ago it was dismissed as an expensive, unreliable irrelevance.

By the last quarter of the twentieth century, the world had exploited apparently inexhaustible and cheap deposits of fossil fuels for more than 200 years, and saw no need for "alternative" sources of energy.

It was a crisis to this oil-dependent world that opens the story of terrestrial solar energy, told in this Part I. It is followed in Part II by fuller descriptions of the pioneering people, events, and organizations, which created this new energy sector. Finally, Part III gives a comprehensive reference section of sources and resources.

The Solar Generation: Childhood and Adolescence of Terrestrial Photovoltaics, First Edition. Philip R. Wolfe.
© 2018 by the Institute of Electrical and Electronic Engineers, Inc. Published 2018 by John Wiley & Sons, Inc.

1

Origins of Terrestrial Solar Power

"Where shall I begin, please your Majesty?" he asked.
"Begin at the beginning," the King said, very gravely, "and go on till
you come to the end: then stop."

Lewis Carroll [3]

If we were to begin at the beginning of the story of solar energy, we would go all the way back to the formation of the sun and the earth. Virtually all the energy we use comes straight from the sun; only atomic and geothermal energy use a resource that is not directly solar in origin. Sunshine fueled the growth of the organisms that gave rise to the earth's coal, oil, and gas deposits. Today it grows the trees and crops for our biomass and biogas production. It is the source of the rain for our hydropower and wind for the turbines.

But let's not begin this story in prehistory.

If we were to begin with when solar energy was first used to produce electricity, we would go back before Albert Einstein's Nobel Prize in 1921, which many will be surprised to hear was for his work on photoelectricity [4], not quantum theory. We would need to look at the previous century's achievements of the Becquerel family in discovering the phenomenon of deriving an electric charge from sunlight – the photovoltaic effect [5]. We'd look at the work of Bell Laboratories and others in the mid-twentieth century on early solar cells, and the first applications of photovoltaic (PV) devices in space in 1958. But I intend to cover the birth of photovoltaics only fleetingly.

Instead, this story begins with the "first oil crisis" of 1973–1974. That one event, more than any other, heightened mankind's awareness that the energy sources it so desperately depended on were neither ubiquitous nor infinite. This led to dramatically increased interest in what at the time was called "alternative energy." It also led to the formation of the International Energy Agency in 1974. Furthermore, the sudden increase in the oil price brought about by the crisis

The Solar Generation: Childhood and Adolescence of Terrestrial Photovoltaics, First Edition. Philip R. Wolfe.
© 2018 by the Institute of Electrical and Electronic Engineers, Inc. Published 2018 by John Wiley & Sons, Inc.

started a progressive change in attitude about the value of energy, and made hitherto costly looking alternatives more attractive.

This congruence of factors led to what I consider to be the start of the terrestrial photovoltaics industry. This is when Joseph Lindmayer, Peter Varadi, Bill Yerkes, and Ishaq Shahryar left the US space solar industry to establish independent PV companies; and Elliot Berman persuaded Exxon to back his solar enterprise. It is when electronics and energy companies in Europe and Japan shifted photovoltaics out of their research laboratories and into business units. And it is when marketing of solar systems for use on earth really began, although there had been isolated earlier applications.

Having decided where to begin, the next decision is where to end. The end of a millennium is a notable juncture in any case, and 1999 was the year when cumulative solar photovoltaic capacity reached its first gigawatt (1 billion watts) [6]. It also proved to be a turning point for renewable energy. It was just into the new millennium when national feed-in tariffs were first introduced for solar power; and they, more than any other mechanism, created the climate for explosive growth in renewables generally and solar power in particular.

This book therefore focuses on terrestrial photovoltaics between 1973 and 1999. For convenience, I refer to this period as simply *our time frame, the early PV era,* or *the first solar generation.*

1.1 OPEC Oil Crisis

The first oil crisis was sparked in October 1973 when some members of OPEC proclaimed an oil embargo in response to American supply of arms to Israel in the Yom Kippur War. At the time petroleum consumption by industrialized countries was rising rapidly and the price of oil was about $3 per barrel.

The Organization of Arab Petroleum Exporting Countries (OAPEC) comprised the Arab members of OPEC including Syria and Egypt, who had started the war. The embargo covered shipments not only to the United States but also to Canada, Japan, the Netherlands, and the United Kingdom.

The resulting strain on international relations led to intensive diplomacy headed by the Nixon administration's Secretary of State, Henry Kissinger. The prospect of a negotiated settlement between Israel and Syria eventually led to the embargo being lifted in March 1974, by which time the global oil price had risen fourfold to almost $12.

OPEC members, led by Saudi Arabia's Sheikh Yamani, recognized the leverage they could exert and agreed to use the world price-setting mechanism for oil to increase their income. The continuing relatively high price of oil, and a keener appreciation of the concept of energy security, led industrialized nations to consider other energy options more actively.

1.2 Energy Security

Before this oil shock, the supply of fossil fuel was assumed to be virtually infinite. It was OPEC's constraint on supply that, albeit artificial, sowed the seeds for a more realistic view.

Although the expression "peak oil" wasn't coined until later, some analysts now started to consider the lifetime of available fossil fuel deposits and the dynamics between rates of discovery, exploitation, and consumption. Shell's M. King Hubbert had first postulated his peak theory in 1956 [7] and projected in 1974 that US oil consumption could by 1995 exceed the pace of new discovery [8]. The name Peak Oil was given to this phenomenon [9] in 2002 and this concept is now widely accepted, although the precise dates remain a topic for debate.

The concept of "energy security" has subsequently been broadened to take into account other factors such as the political stability of the regions where energy is produced, and risks associated with transporting it to the point consumption. Other threats such as terrorism also need to be weighed in the balance.

The second oil crisis provided further impetus to the growth of renewable energy. This started in 1979 when oil production in Iran declined after the revolution there, and it was exacerbated the following year by the outbreak of the Iran–Iraq war, which almost stopped production in Iran and severely curtailed Iraq's output.

So energy security, in the form of the availability and price shocks of the oil crises, provided the first major stimulus for terrestrial photovoltaics; and although the circumstances have changed, energy security remains a substantial driver today.

Let's briefly consider other significant drivers, even if their impact was not so weighty at the start of our time frame.

1.3 Climate Change

Climate science was still very much in its infancy in the early 1970s. The link between atmospheric carbon dioxide and methane with global temperatures was not widely recognized; or as a skeptic might put it, "climate change had not yet been invented."

Climate change considerations did not in practice become a substantial inhibitor to fossil fuel usage, or a main driver to the growth of renewables, until some two decades later. Toward the end of the twentieth century, climate change became the primary motivation for supporting renewables and creating the incentives that allowed solar energy to progress from adolescence toward adulthood.

1.4 Other Drivers of the Early Renewable Energy Sector

Along with the external drivers, summarized above, a number of internal drivers were also at play.

Key to the development of any new industry are the companies and people who get it started. From the very beginning, the terrestrial PV space was occupied by a broad cross section of independent and multinational companies. For the independents, we can presume that they were inspired by the concerns of their principals.

The motivation of the larger companies is not always so clear-cut. For many, it was an issue of strategic diversification or expansion, as further discussed in Chapter 7. But some seem to have viewed their involvement as market research, to keep tabs on this new sector – maybe even seeing it as a threat – and others seem to have seen it as a PR exercise.

Ultimately, of course, any new industry is all about the people. Tales of many intrepid PV pioneers will crop up during this story, and some of them are individually profiled in Chapter 10.

1.5 That Sisyphus[1] Feeling

Terrestrial usage of solar power was viewed with widespread skepticism for several decades. This resistance took many forms.

For a start, many people just did not believe that it could work. They thought it was some kind of trick; "you just can't produce energy out of thin air." This may be partly because PV is a solid-state technology – harder to understand, when all previous experience of electricity generation was based on rotating machinery.

Next came questions about reliability and longevity. Of course, any new invention needs to prove itself, and people want to see it in action before they commit. The whole field of semiconductor technology was only a few years old and people weren't yet used to electronic devices, which are now so much part of everyday life.

More rational resistance came on the grounds of cost, and this will be addressed further as the book progresses. Remember that most traditional power generation involves plants where the upfront capital cost is relatively low, but the operating and fuel costs are high. Neither power engineers nor finance directors had economic models that could attribute fair value to the negligible running costs of a solar power plant.

Resistance was often particularly marked in more temperate climates. "Power from the sun? You've come to the wrong place!" Many assumed that PV

---

1 Mythical Greek king condemned by the Gods to eternally roll a boulder uphill.

technologies could work only in the sunniest places. It seemed incongruous that daylight is the only input required, and counterintuitive that systems are more efficient at low temperatures.

Finally, it doesn't take a conspiracy theorist to recognize that a successful solar industry threatens the business model of traditional energy producers. Political pressure was certainly brought to bear to slow the deployment of many environmental technologies. There are those who believe that similar tactics were adopted inside the industry as well.

All in all, there was a lot of resistance, and even more inertia, to slow the early progress of the industry. The effort required to secure and maintain the attention, political support, and funding felt like the unending struggle to push a boulder uphill.

2

What Is Photovoltaics?

Low value, expensive, unreliable, high capital cost, land hungry, intermittent energy
  Solar and wind simply don't work, not here, not anywhere.
                                           Australian politician Steve de Lacy [10]

This quotation so admirably encapsulates many of the attitudes prevalent in the early PV era; it justifies inclusion, even though it is much more recent.

Solar power is so familiar to us now that it is worth remembering the absolute wonder that would have been felt on first seeing a solar cell in operation. At a time when our electricity came from huge smoke-belching power stations, it was spectacular to connect a small flat gray disk to a fan and see the blades turn. Shadow the disk with your hand and the fan stops again! At the time it seemed like magic. It may be the memory of this first miracle that captivated so many of the early PV pioneers for the rest of their working lives.

Chapter 1 selected our time frame – the 27-year period from 1973 to 1999 – as PV's "childhood and adolescence." I promised to cover the earlier period only fleetingly,[1] so here goes.

2.1   Prequel – The Birth and Infancy of Photovoltaics

The first published observation [5] of the photovoltaic effect was by 19-year-old French scientist Alexandre-Edmond Becquerel in 1839, possibly working with his father, the physicist Antoine César. Becquerel noticed that when light strikes certain substances, an electric charge accumulates. It seems strange to us today that they did not do more with this discovery, but the field of scientific endeavor at that time was a blanker canvas, and the talented Becquerels had so many fresh areas to explore.

---

1 John Perlin's book [13] is commended to readers wanting the fuller story of this period.

The Solar Generation: Childhood and Adolescence of Terrestrial Photovoltaics, First Edition. Philip R. Wolfe.
© 2018 by the Institute of Electrical and Electronic Engineers, Inc. Published 2018 by John Wiley & Sons, Inc.

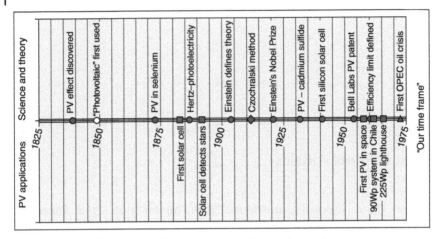

Figure 2.1 Simplified timeline of earlier photovoltaics development.

The word "photovoltaic" was coined in England a decade later by Alfred Smee [11]. The word – not quite interchangeable – "photoelectric" came into use following Heinrich Hertz's 1887 discovery that ultraviolet light altered the voltage needed to produce a spark between electrodes [12].

In 1892, the Irish mathematician, George Minchin produced his own selenium photovoltaic cells for astronomical measurement of light from stars, although the first practical solar cell had been fabricated almost a decade earlier by the American inventor Charles Fritts.

Most scientists, however, were skeptical about the photovoltaic effect, because it didn't fit with the understanding at the time of light simply as a wave; until Albert Einstein applied his knowledge of quantum theory to this field [4]. In fact, his Nobel Prize in 1921 was awarded "especially for his discovery of the law of the photoelectric effect." Meanwhile, various researchers discerned the PV effect in other materials [14]. Despite all this fine science, Becquerel's discovery went largely unexploited as a source of power for more than a century, as shown in Fig. 2.1.[2]

This lack of practical interest was in part due to the low conversion efficiency of these early experiments, where the electrical energy produced was often less than 1% of the incident light energy. As late as 1949, Vladimir Zworykin and Edward Ramberg concluded [15] that "It must be left to the future whether the discovery of materially more efficient cells will reopen the possibility of harnessing solar energy for useful purposes."

---

2 This also shows Jan Czochralski's breakthrough in the production of monocrystalline silicon – a method relevant beyond solely the solar sector.

The first progress in this respect came in the 1950s, when research teams at the electronics company RCA and at AT&T's Bell Laboratories developed working PV solar cells with up to 8% efficiency. In 1954, Daryl Chapin, Calvin Fuller, and Gerald Pearson at Bell Labs developed [16] a "Solar Energy Converting Apparatus," submitting a patent application [17] that was in due course granted.

Meanwhile, participants in the embryonic space program were aware that missions would be severely restricted if they had to take all their energy supplies with them in the form of batteries. The US Signals Corps' William Cherry encouraged RCA to work on solar cells and in 1958 the Vanguard I satellite [18] was the first practical application of PV, with less than 1 W of capacity. Later that year, Explorer III, Vanguard II, and Sputnik-3 all carried PV-powered systems. Early cells were small crystalline silicon devices, typically sized at just 1 cm square. Cost was not a primary issue; indeed, it was typically around $1000/W of capacity.

With very few exceptions [19], nobody before the first oil crisis gave any serious consideration to PV for terrestrial use, although Philips, Hoffman Electronics and Sharp all undertook early installations.

2.2 Where Does the Energy Come From?

Solar energy is electromagnetic radiation emitted by the sun. Enough of it falls on the surface of the earth in 1 h to power it for a whole year. It has been estimated [20] that it would require about half a million square kilometers of solar systems to provide all the energy we need. This is about the area of Uzbekistan – although Uzbek readers should note that I say this only for comparison purposes, not as a proposition! Every location on earth receives sunlight for at least part of the year. The amount of solar radiation that reaches any one spot on the earth's surface varies with location, season, time of day, weather, landscape, and shading.

The sun strikes the earth's surface at different angles, ranging from just above the horizon (Fig. 2.2) to 90° directly overhead. The lower the sun is in the sky, the farther its rays have to travel through the atmosphere, becoming more scattered and diffuse. This distance through the atmosphere – or strictly the ratio of the distance to that when the sun is directly overhead – is called the "air mass."

The tilted axis of the earth's rotation affects the amount of sunlight at a particular location and time of year. Locations at around 40° of latitude receive more than twice as much solar energy in midsummer than midwinter, and this ratio increases to about 5:1 around 56° of latitude (Moscow, Edinburgh, or Cape Horn). In the tropics, summer-to-winter variations are smaller.

As the sun's rays travel further through the atmosphere, some are absorbed, scattered, and reflected by clouds, mist, and water vapor, sand, dust, smoke, and

Figure 2.2 PV anywhere on earth – even Antarctica [21] where the modules are vertical.

other pollutants, and of course the air itself. The solar radiation that reaches the earth's surface straight from the sun without being diffused is called direct radiation. The light that has been scattered is called diffuse radiation and accounts for most light on cloudy days. The sum of the diffuse and direct solar radiation is called global radiation.

2.3 The Photovoltaic Effect

No generating station actually creates energy; it converts it from an existing energy source into a different form. We extract oil from the earth's surface so vehicle engines can convert its chemical energy to motive power. We mine radioactive isotopes so nuclear power stations can convert atomic energy to heat and thence power. Now we can harvest sunlight by using solar cells to convert light energy into electricity (Fig. 2.3).

The solar cell is the primary active component of a solar power system. It is supplemented by other subsystems to condition and transport the energy and sometimes to store it. A solar cell produces electricity when the photons of light pass energy to negatively charged electrons in its semiconductor material,

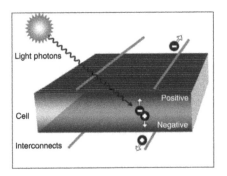

Figure 2.3 Photovoltaic effect in a solar cell.

allowing them to break free. The residual atom is then left with a net positive charge.

Because of the way solar cells are processed, the negative charges are attracted to one surface and the positive charges to the other. Conductive grids on the front and back surfaces of the solar cell collect these positive and negative charges. These contacts are in turn connected to wires to conduct the flow of direct current (DC) from the solar cell to power an external circuit.

We won't explore the technical characteristics in detail, but this summary helps understand how the PV applications described in the next chapter evolved.

2.3.1 Photovoltaic Operating Characteristics

The so-called current–voltage (or I–V) characteristic, illustrated in Fig. 2.4, shows how much current and voltage can be delivered by a solar cell at different levels of light intensity. The cell can operate at any point on the line, and the actual output therefore depends on the external circuit to which the cell is connected.

For example, the extreme right-hand end of the line shows the so-called open-circuit voltage (V_{OC}) – the potential that exists between the contacts at the front and back of the solar cell, when they are not connected to any external circuit and are therefore not delivering any current.

The left-hand end of the line shows the short-circuit current (I_{SC}), which flows if the solar cell contacts are directly connected together – that is, at a zero voltage. Note how solar cells are resilient and can be short-circuited without damage – don't try this with most other electricity generators!

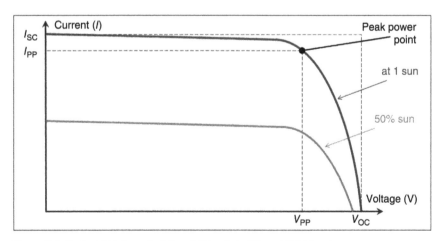

Figure 2.4 Typical I–V characteristics of PV cell at different light levels.

Any electrical engineer can tell you that power is the multiple of current and voltage. Solar cells therefore produce no power either at open circuit (where current is zero) or at short circuit (where voltage is zero). The point at which power is at its maximum is called the peak power point. This is also shown on the characteristic. The output at this point, known as "watts peak" (W_P) is the key parameter against which cells and modules are rated. As the name implies, there is a benefit from operating solar systems as close to this point as possible. The capacity of a PV cell, module, or system, quoted in W_P (or kW_P, MW_P etc.) is a rating of the power it delivers when operating at the peak power point under standard test conditions, explained in Section 12.3.

A measure of the electrical quality of the cell is called the "fill factor," which is the ratio of the peak power to the product of the open-circuit voltage with the short-circuit current (i.e., the ratio of the areas of the two dashed rectangles shown in Fig. 2.4). The "squarer" the I–V curve, the higher the fill factor.

The point on the characteristic at which the solar cell will operate in practice will depend on the external circuit to which it is connected. That circuit also has an I–V characteristic; and the whole network will operate where the two characteristics meet. By way of example, a battery at a given state of charge operates at a fixed voltage, that is, a vertical line on the characteristic. The current with which the solar cell would charge that battery therefore varies with the intensity of light to which the cell is exposed.

There are many factors that affect the performance of solar cells of different types, and this book does not intend to address the science of these, although we touch on "bandgap" in Sections 4.4 and 4.7. Readers should look elsewhere [22] if they want a better understanding of this and parameters such as the "minority carrier lifetime" and "surface recombination."

It is important to note that the PV effect is driven by the level of incident light energy. Solar cells therefore work whenever there is daylight, even on cloudy days. They do not need heat, and in fact work more efficiently at low temperatures.

2.3.2 Solar Cell Production

Here is a thumbnail sketch of how monocrystalline silicon solar cells are traditionally produced to achieve this clever task (a fuller account of solar cell technology is given in Chapter 3).

Monocrystalline silicon ingots are grown by introducing a small crystal at the top of a bath of pure molten silicon. As this is slowly rotated and pulled upward, the silicon solidifies in a sausage shape, with all its molecules aligned in a coherent crystal lattice. This is called the Czochralski process after the Polish chemist Jan Czochralski, who invented this method for producing crystalline metals in 1916 [23].

The ingot is "salami-sliced" into individual thin wafers, and these are placed in ovens, where small amounts of two chemicals (often boron and phosphorus) diffuse into the two faces to selectively attract the positive and negative charges as described above.

Finally, conductive contacts are attached to the front and rear surfaces to collect the current produced. The cells are then usually connected together in strings and packaged for weather resistance and safety in solar panels, known as "solar modules."

2.4 From Theory to Practice: Applying PV Technology

In hindsight, it may seem surprising that several decades elapsed after the discoveries at Bell Laboratories in the 1950s, before any widespread use of solar electricity. There are several reasons.

Scientifically, this was a very exciting era, especially because of dramatic progress in semiconductor technology. Many of the laboratories active in photovoltaics were also researching diodes, transistors, and other devices with more immediate commercial applications; so developments in the PV sector were often overwhelmed by the larger semiconductor industry.

At the Philips Physics Laboratory in the Strijp suburb of Eindhoven in 1955, for example, 19-year old apprentice Kees Ouwens was employed for making diodes and transistors. He sawed silicon rods into wafers, diffusing them as semiconductors and applying contacts. In the spring of 1956, he made a few solar cells in the same way, and had soon produced what was probably the first solar-powered transistor radio. But the next year he was back on the "day job" posted to the Philips UK subsidiary Mullard in the production of transistors.

Second, energy prices in the 1950s were extremely low. Although PV had a myriad of potential applications, most of these were already served by cheap alternatives based on fossil fuels. Except in space, where there were few other options, PV looked like an expensive alternative to existing energy sources.

There was therefore very little terrestrial deployment of photovoltaics, although a few pioneers did install development systems to see how they worked. The earliest such installations (of which I am aware) were for isolated mining sites in Chile and lighthouses and navigational buoys in Japan. Other off-grid and remote applications gradually started to emerge, as further described in Section 3.2.

Of course, the whole concept of photovoltaics was entirely foreign to the first prospective customers. Early marketers made solar gadgets – often little rotating fans, which they could give away to show that PV actually works. Lucas fabricated elegant mahogany boxes with two Solar Power Corporation P1002 modules in the lid, powering a fan and a light and some meters.

Arco Solar[qv] invited customers to demonstrations at hotels where solar panels powered a pump in the swimming pool. When the clients walked in front of the panels, the pump would stop; when they moved away, it would start again. "After they did that a couple of times," says Peter Aschenbrenner [24], "you could 'see the light go on.'"

2.5 What Is a Solar PV Energy System?

The photovoltaic cells described above are the active parts of a solar generator – which converts light energy into electricity. To meet the needs of energy users, other components are added to make a complete generating system.

2.5.1 Non-PV Solar Power Systems

Most, but not all, solar energy systems are based on photovoltaic cells, and when I use "solar power" or "solar energy" in this book, I am referring to PV. For the record, however, let's quickly mention the other major forms of solar energy conversion.

Concentrated solar power (CSP) uses the sun's heat to generate electricity. One of the two main methods for collecting heat at high temperatures for this application is to use so-called "power towers." A large field of mirrors focus the sun's rays onto a boiler at the top of a tower, as illustrated in Fig. 2.5.

An alternative CSP approach uses parabolic reflector "troughs," which focus the sun's rays onto a tube mounted at the focal point. The heat is collected in a transfer fluid circulated through these tubes. The SEGS (solar electric generating system) projects built between 1983 and 1990 at three sites in the Mojave Desert comprised nine solar power plants using parabolic trough collectors (Fig. 2.6).

Whichever thermal collector technology is adopted by these CSP systems, the heat then drives a steam turbine, similar to those used in traditional power stations. The early business of Flachglas Solartechnik[qv] was the production of mirrors for concentrated solar power systems.

Figure 2.5 Not PV: concentrated solar "power tower" in Spain [25] at the Gema-solar project.

Figure 2.6 Not PV: mirror troughs collect solar heat at SEGS [25].

Thermal systems can also be used for providing solar hot water (SHW) and space heating for buildings. Solar thermal panels collect the sun's heat using either black collectors or evacuated tubes and this heat is then distributed through water, air, or other transfer fluids to the building's hot water or heating system.

2.5.2 Solar System Configuration

Having briefly side-tracked into thermal solar energy conversion, we return to our main topic – photovoltaic systems that convert light directly into electricity.

There are two main configurations: one for stand-alone systems, using DC power, and the other for AC applications, often connected to the grid. These alternatives are illustrated schematically in Fig. 2.7, and the main subsystems of each are outlined in the remainder of this chapter.

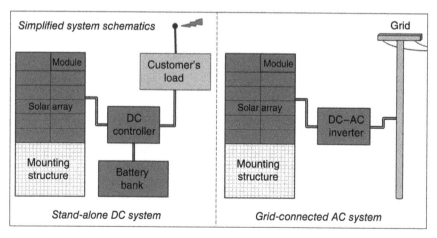

Figure 2.7 Simplified PV system schematics for DC and AC applications.

Figure 2.8 Monocrystalline silicon solar cell [26].

2.5.3 Solar Cells

Solar cells are the core of a PV system, responsible for converting incoming light into electrical energy. Most solar cells are made from crystalline silicon wafers (Fig. 2.8), although other technologies have been developed, as discussed in Sections 4.4 and 4.6. In principle, all cells operate similarly, although the characteristics do vary for different materials.

Solar cells become electrically active, when exposed to light. The voltage each cell delivers is a function of its material properties, typically on the order of ½ volt. Above a threshold of about 10% sun,[3] this voltage is broadly independent of the radiation level, as shown in the I–V characteristic we saw in Fig. 2.4. The current produced by each solar cell is roughly proportional to the light level and to the area of the solar cell.

Solar cells can be connected together in both series and parallel. In accordance with basic electrical theory, the overall voltage is proportional to the number of series-connected solar cells. Multiple parallel connections increase the overall current. In both cases, unproductive losses can be introduced if cells interconnected in this way are not well matched.

2.5.4 Solar Modules

Solar cells are active electrical components, needing protection to prolong their life; so they are usually encapsulated in packages known as solar modules. The solar module has evolved over time as further described in Section 4.7, but the most common configuration comprises a front cover of toughened glass behind which the solar cells are mounted in some form of encapsulant and protected by a weatherproof rear membrane.

---

3 The light level colloquially known as 1 sun is defined by international standards and is equivalent to unshaded sunlight at a subtropical location – see Section 12.3.

Figure 2.9 A single solar module with 36 round cells [27] powers this phone box in the Middle East.

Some types of thin film cell, as further described in Section 4.4, are deposited directly onto the glass cover sheet as a continuous film. In this case, unlike crystalline silicon wafers, the solar cell has no independent existence outside the module.

Apart from providing mechanical protection, the solar module allows a number of cells to be interconnected to provide a suitable combination of current and voltage. Because most early systems were designed to charge nominal 12 V batteries, mainstream solar module manufacturers rapidly migrated to standard products using around 36 series-connected cells (Fig. 2.9). This gives a nominal output of about 18 V, suitable after temperature and cable losses, for charging lead–acid batteries even as they approach full charge at around 14 V.

Solar modules typically also incorporate a frame for fixing to a mounting structure and a connection box or output cable.

2.5.5 Balance-of-System

Throughout the early PV era the cost of solar cells was relatively high, making solar modules the dominant element of a PV system. The structures, electronic controls and batteries all together rarely accounted for more than 30% of the total cost. All these other elements were therefore lumped together under the term "balance-of-system" (BOS).

Let's look briefly at each of these subsystems.

2.5.6 Array Structures and Trackers

Collections of solar modules are normally mounted on structures to optimize their orientation for incoming solar radiation (Fig. 2.10). In the early PV era,

Figure 2.10 Solar modules on a typical "ground mount" structure [27] by the radio tower they power.

most of these structures were static at a fixed angle toward the South in the northern hemisphere and North in the southern hemisphere.

The system designer aims to calculate the optimum tilt angle based on local solar radiation and the required energy output. In stand-alone DC systems, this leads to a steeper angle to maximize winter output, thus giving more uniform performance throughout the year. Grid-linked systems, on the other hand, often use a shallow angle to maximize summer output and so overall energy production.

A minority of systems use trackers (see Fig. 2.12), so the solar arrays follow the sun across the sky. The benefit of tracking in terms of the higher energy output is partially offset by the energy required to operate the trackers, and the relatively lower reliability of a moving system. The relative benefits are greater in sub-tropical climates with high levels of direct sunlight. More diffuse radiation, due to cloud cover in temperate climates, make trackers less appropriate here.

2.5.7 Concentrator Systems

One approach to reducing system cost is to focus the incoming sunlight, using either mirrors or lenses. This means that the area of expensive solar cells can be smaller than the total collector area. A wide range of concentration ratios have been attempted, from multiples of 2 or 3 all the way up to 1000 suns and beyond.

High concentration ratios usually use Fresnel lenses to focus the light onto the solar cells, such as that illustrated in Fig. 2.11.

Arco Solar's 1983 Carrisa Plain project used two mirrors on either side of the solar arrays to achieve a concentration ratio of about 3× (Fig. 2.12). However, they found this design did not give the uniformity of illumination needed to achieve the higher output that had been predicted. In the end only two-thirds of the project used these concentrators; the balance reverted to standard flat plate systems.

Concentrator systems need double-axis trackers to follow the sun, and are unsuitable for areas with high levels of diffuse radiation. This approach was not widely used during our time frame, and seems ever less applicable as solar cell prices fall.

Figure 2.11 Fresnel lens concentrator [28].

Figure 2.12 Low concentration tracking array [25] at Arco's Carrisa Plain project.

2.5.8 Batteries

Having covered the main mechanical subsystems, let's move on to the electrical configuration. Batteries are used in stand-alone DC systems, the most common application during the early PV era. Grid-connected AC systems, on the other hand, tend to use the electricity grid instead of storage, although latterly some now also incorporate batteries.

Most of the batteries used at the time were standard lead–acid cells, originally designed for automotive, marine, or forklift truck use. The batteries are typically the shortest-life subsystem within a solar generator installation. In good systems, the battery should give a lifetime of maybe 10 years. Furthermore, most early lead–acid batteries required servicing and electrolyte replenishment typically every 6 months.

Sealed and gel cell batteries (see Fig. 4.17) that do not need electrolyte replenishment became more prevalent during the 1990s.

Lead–acid batteries give off acidic vapor and operate better if not too hot; so they are best housed in a ventilated battery room, if there is a building on site. Alternatively, they can be placed in vented battery enclosures, located under the solar array for shading (see Fig. 2.13 and Fig. 3.4).

Figure 2.13 Heli-lifting batteries in enclosure to mountain-top site [27].

2.5.9 Electronic System Controllers

Most early PV systems were used in high-reliability professional applications detailed in Section 3.2. They employed control electronics to optimize battery charging and to protect the overall system (Fig. 2.14).

The first requirement of the electronic controls is to condition the charge and discharge of the batteries. Battery life is enhanced by ensuring that it is not over charged nor over discharged. Controllers therefore incorporate a regulator that disconnects the solar array when the battery reaches full charge. Many also incorporate a low-voltage disconnect to isolate the load equipment as the battery nears full discharge. Specialist applications such as cathodic protection require further output controls.

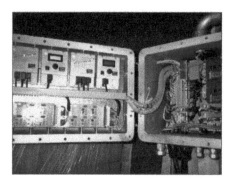

Figure 2.14 Electronic DC system controller in explosion-proof container [27] for use on offshore oil platform.

Electronic controls became increasingly sophisticated over time with circuits designed to hold the solar arrays near their peak power point. Many incorporated telemetry, so that users could check on the performance of their solar power systems without having to visit each site.

2.5.10 Inverters

Solar systems designed to deliver AC power need inverters to transform the direct current produced by the solar arrays into alternating current. In cases where these systems are connected to an existing electricity grid, the inverter also needs to synchronize with the network.

Initially, the industry used standard inverters, but these proved unreliable and inefficient because of the wide operating range of PV generation systems. In 1983, for example, Raju Yenamandra [29] bought 2 MW of inverters for his first 1 MW project in the hope that 50% would work.

Inverters were soon developed specifically for solar applications by SMA[qv] and others. An early enhancement was to incorporate maximum power point trackers (MPPT) into inverters, so that the array operated close to its peak power point irrespective of the operating conditions. Large systems can use a few big so-called "centralized" inverters or a larger number of "string inverters," one for each solar subarray.

Solar inverter technology has continued to advance, with many models able to operate at efficiencies in the high 90% range over a wide range of operating conditions.

3

Terrestrial Solar Applications

". . . a fringe market that appealed to a faction of eco-hippies"
Jacquie Ottman [30]

Having read this far, you know that photovoltaic power works, it is highly reliable, and is getting cheaper. But trying to explain just these points to early potential customers proved so difficult in the infancy of the PV era.

As to any new technology, there were a number of objections, ranging from the educated ("Surely that contradicts the First Law of Thermo-dynamics"[1]) to the outlandish ("Won't that use up all the sun's energy?"). These doubts led many to see photovoltaics as no more than a fringe or niche business.

The industry today is indebted to the early adopters who saw past these objections. These pioneers got excited when the solar system really worked, reliability was up, operating costs were down, and their risk had paid off. They told their friends and colleagues, and step-by-step the word spread.

A photovoltaic system is simply a generator. It can be used wherever electrical power is needed. Furthermore, because it is modular all the way down to individual solar cells, it can produce power at any scale. A single solar cell can replace, or be coupled with, a battery to energize the smallest electronic device. Solar modules can be connected together in their millions to replace the largest coal-fired power stations.

The range of possible applications for terrestrial PV is therefore endless. This section looks at key drivers and then the main types of application, for which terrestrial PV has been deployed, in roughly the order in which each market materialized.

---

1 Energy cannot be created or destroyed (to paraphrase).

The Solar Generation: Childhood and Adolescence of Terrestrial Photovoltaics, First Edition. Philip R. Wolfe.
© 2018 by the Institute of Electrical and Electronic Engineers, Inc. Published 2018 by John Wiley & Sons, Inc.

3.1 Energy Comparatives

A factor that affects the deployment of any new technology is the extent to which other alternatives exist, and how it compares to any such other options.

Photovoltaics, as a power conversion technology, must be measured against other electricity generation sources. A key comparator is the relative cost, to be discussed in Chapter 6. There are also nonfinancial differentiators that proved crucial in the early years of the PV industry.

Most other forms of electricity generation rely on rotating machinery, powered by steam from a boiler, internal combustion engines, or a mechanical source such as a wind or water flow. Photovoltaics, by contrast, is a solid-state phenomenon where the only moving parts are the electrons. This means that it is not only silent but also more reliable than competing energy technologies, because there are no mechanical components to wear out or maintain.

The next consideration is the input energy source. On earth, sunlight shines almost everywhere at some time of the day. The wind too blows all over the earth's surface, although in many places not consistently enough to be viable. Hydropower, wave, and tidal resources are yet more geographically restricted. Fossil fuels can in theory be transported almost anywhere for thermal generation, but in practice many locations are inaccessible.

Away from the earth's surface, of course, things change. In space, for example, solar power is pretty much the only available resource. Under ground you can forget it.

The efficiency of conversion – early solar modules converted about 10% of the incident solar radiation – seems at first sight to be a relative drawback. However, this looks less significant, when you realize that most other energy vectors also derive their input from the sun, and typically with even lower end-to-end efficiency. It has been calculated, for example, that when you look at the conversion of sunlight to plant and animal matter – to make the fossil fuel deposits to burn to generate steam to power a closed-circuit gas turbine, the end-to-end efficiency is on the order of 0.5%.

All in all, area efficiency is not a major constraint. Even at the modest conversion efficiencies of early solar cells, the area required to generate all of the world's energy was only about half a million square kilometers [31]. Enough solar radiation falls on the surface of the earth in 1 h to power it through photovoltaics for a whole year

A more significant comparative constraint is the timing and availability of the energy resource. Sunlight levels deliver an annual average of about 6 h of nominal output per day in the best subtropical locations, and less than 3 h per day in temperate regions. This equates to an availability factor between 10 and 25%, depending on the location. This is of a similar order to wind power, but less than a good hydroelectric site. Fossil- and biomass-powered thermal generating stations can deliver electricity essentially on demand, and with availability factors better than 70%.

An additional dynamic is the scale at which energy is needed, and can be produced. Solar power is modular and almost infinitely scalable. It works with fundamentally the same efficiency from fractions of a watt to gigawatts. No other energy source is this versatile. Thermal power stations are typically viable only in the hundred megawatt to gigawatts range. Hydropower can cover the range between tens of kilowatts to gigawatts; and wind turbines are normally rated between a few kilowatts and a few megawatts, beyond which they can be clustered together in windfarms.

Finally, when one considers the sustainability of different options, solar power – like wind, hydro, wave, and tidal – delivers energy with essentially no emissions. By contrast, coal-, oil-, and gas-fired power stations and diesel generator sets emit significant levels of carbon dioxide. Atomic power stations also produce harmful by-products, in this case in the form of nuclear waste.

It is the combination of these relative technical factors, together with the – initially rather unfavorable – comparative economics, that defined the early markets for terrestrial PV.

3.2 Professional Stand-Alone Systems

In light of the competitive situation, described above, the most propitious applications are those where existing energy costs are high, power requirement is low, and the practical benefits of PV, such as reliability, are most prized.

Replacement of batteries and diesel generator sets in professional off-grid applications offered the first commercial market. Primary batteries are relatively expensive, and must be replaced each time they run out, while diesel generators can be unreliable and need regular refueling and maintenance.

The main stand-alone professional applications for solar power are in the transport, telecommunications, and petroleum sectors; and we will look at a few examples in this chapter.

3.2.1 Transport and Navigational Uses

This section deals with stationary installations in the transport sector. Mobile applications – PV-powered cars, boats, and planes – are covered in Section 3.8, and we have already noted in Section 2.1 that the earliest applications for PV were in space vehicles where it remains the primary energy source to date.

Figure 3.1 Solar array powering Mumbles lighthouse in Wales.

Terrestrial applications in the transport sector started before the first oil crisis. One of the earliest recorded practical uses of solar cells on earth was for navigational aids in Japan, where the Ogami lighthouse was powered by a solar system from Sharp[qv] in 1966 (see Fig. 11.18). In the early PV era, navigational aids represented a continuing commercial market for the first terrestrial PV companies (Fig. 3.1).

The navigational beacons along the North-West Passage in Canada, for example, are redundant in winter, when the seaway is ice-locked, but are needed again in the spring. The Canadian Coast Guard used to power these with primary batteries, which needed replacing every year – an expensive and difficult task, especially at a time when the ice was just starting to melt. Canadian Coast Guard's Don Gifford undertook extensive testing of photovoltaics to prove its reliability, and then replaced his primary batteries with solar power systems. These include electronic controls to switch the beacons off during the winter, when they are not required, protecting the solar storage batteries. In spring, the solar systems automatically wake up and switch the beacons back on. The higher levels of sunlight keep the navigational aids going throughout the summer.

Many other coastguard departments around the world adopted a similar approach. The US Coast Guard, Trinity House in the United Kingdom, the Japanese Coastguard, and others soon became experts in the deployment of PV navigational aids, as did specialist suppliers such as Automatic Power, Orga, and Tideland Signal. Most DC navigational aids used 6 or 12 volt batteries, so the PV systems were configured to these voltages (Fig. 3.2).

Railway track circuits, which sense the passage of trains for signaling purposes, similarly need small amounts of power in remote locations with high reliability. This application provided another early market for the PV industry, especially in the wide open spaces of North America, where CP and CN Rail were among the early adopters. System voltages for this application, too, were 12 V or less – some as low as 2 V.

Figure 3.2 Buoy with R&S^{qv} solar-powered nav-aid [25].

3.2.2 Telecommunications Applications

"At present applications appear possible in telephone, telegraph, radio and television – where the power need is small but often in remote locations where no power lines are located."

Gordon Raisbeck, Bell Laboratories [32]
(1955 – the year after Bell's solar cell patent application)

Another sector needing reliable power, often in remote locations, is telecommunications. Again, the first terrestrial solar-powered telecoms applications started even before our time frame; with space cells from Hoffmann Electronics being used for a trans-continental microwave radio broadcast by the US Signals Corps in 1960 (see Fig. 10.7).

The national oil company in Oman, PDO, has its head office near the capital Muscat in the north, but most of the oilfields down around Salalah in the south. The communications link between these two areas is crucial, using a network of line-of-sight microwave repeaters located on mountain tops a few kilometers from the coastline (Fig. 3.3). In the past, these repeaters were powered by diesel generator sets, which had to be maintained and refueled regularly, all by helicopter. PDO decided in the late 1970s to power the stations from photovoltaics instead. This saved them tens of thousands of dollars per annum. More importantly, it reduced the downtime of the communications network from 4 h in the year before the diesels were replaced to 1.5 min the following year (none attributable to failures in the solar power systems) [33]. When you're running a major oil business, 4 h of lost production is very costly!

Figure 3.3 Mountain-top solar-powered microwave repeater in Oman [27].

There are many different telecommunications systems alongside microwave repeaters, and almost all have migrated to solar power at sites where a reliable electricity supply is not otherwise available.

As a side effect of this change, suppliers of the most efficient telecommunications equipment found they have a competitive advantage. Solar power systems are designed to deliver the power consumption actually required (as discussed in Section 4.10). More power-efficient radios need smaller solar systems, and the overall cost is therefore lower. Equipment manufacturers such as EB Nera, GTE Lenkurt, Harris Farinon, NEC, Sirti, and Telletra benefited from this in the 1980s and 1990s.

Most DC radio equipment is designed to operate at 12 V or multiples thereof (commonly 24 V, occasionally 48 V). The importance of the telecoms market – and the emergence of rural systems charging car and truck batteries – led the early PV industry to standardize on nominal 12 V solar modules.

3.2.3 Pipeline and Oilfield Applications

The petroleum sector was not only a supporter of the early PV industry (see Section 7.2), it was also a regular customer. There are various requirements for reliable power systems in oil and gas fields and for pipelines

The main application in the sector is for the so-called "cathodic protection" of pipelines, wellheads, and other structures. This involves applying a small voltage to the pipeline to prevent corrosion. In the Middle East and other regions, where pipelines extend over long distances of barren and unpopulated land, solar cathodic protection systems were installed every few kilometers to extend pipeline life.

Figure 3.4 Pipeline protection, Libya [27].

Figure 3.5 Offshore platform with solar arrays round the heli-deck [27] being built in Ulsan, Korea.

Cathodic protection is similarly used for other buried parts of structures, such as bridges; so this application is not exclusive to the oil and gas industry. The voltage and current required vary depending on soil conditions, the materials, and age and area of the structure. Most systems are designed to generate power at multiples of 12 V, delivering the output through a voltage converter to give the potential required at the time. These output controls are the small darker boxes under the array in Fig. 3.4.

Another major use for solar by oil and gas exploration companies was for their offshore installations. Offshore platforms need navigational lights and fog horns. They also use telemetry to transmit data about their operations back to base. On unmanned platforms, with no other electricity source, solar systems are often used to power all these devices (Fig. 3.5).

3.2.4 Other Professional Applications

Other applications too began even before our time frame. In 1961 Philips RTC supplied a system producing 26.5 A at 3.3 V for a copper electrolysis experiment in Chile. In those days, the solar cells had efficiencies of only about 6% (Fig. 3.6).

Figure 3.6 Philips 1961 Chile array [25].

There are many other sectors where primary batteries and diesel generators have progressively been replaced by solar systems as a more economic and reliable source of power.

The range of professional applications has continued to multiply subsequently. It is commonplace to see PV modules powering emergency telephones, railway signals, road traffic signs, advertising hoardings, and city center parking meters.

3.2.5 Installing Remote Systems

The preponderance of remote locations for early PV systems made life interesting for installation and commissioning engineers. Unlike today, where systems are typically assembled on site, most professional systems were preassembled as far as possible in the factory, to reduce the time spent in these remote locations. Solar modules, for example, were mounted together and preconnected in subarrays, typically comprising six or eight solar modules.

Small systems could often be installed within a day, but larger systems needed the installation team to be on site for several days. At sites accessed by helicopter (like the mountain tops in Oman), the team would be airlifted to site and left to sleep out in tents until the job was done, and the choppers returned.

Tents might be fine in sunny Oman, where there are few predators (Fig. 3.7). Bob Johnson [34] recalls a project in the forests of New Mexico, where his team spent an uncomfortable night in an old supermarket freezer cabinet to avoid the bears!

Figure 3.7 Helicopter at hill-top site in Oman [27].

3.3 Off-Grid Rural and Residential Development

Professional applications, as already described, provided the first commercial markets for the emerging PV industry, because the customers could afford the relatively high cost and could justify it against improved reliability and reduced operating costs.

Solar generation offers similar benefits in off-grid rural communities, where it can replace expensive – and often dangerous – energy sources such as kerosene and diesel. But because the customers in this sector were not often able to meet the cost of solar systems, much of the early deployment here was funded by aid and grants, often overseen by consultants.

Off-grid rural applications cover in principle everything for which a remote community might require electricity. We will highlight here just the four applications that were most prevalent during our time frame.

3.3.1 Pumping

Before the advent of solar energy, water pumping in the developing world was often done by mule, or by hand. Although both wind- and diesel-powered pumps were available, these were notably unreliable.

Frankly, some of the early solar pumps were too! However, the sector soon benefited from a major development effort by engineering consultancies, pump producers, and the PV companies themselves. More efficient and reliable solar pumps soon became available, and different types evolved to suit various applications including irrigation for crops and livestock, and drinking water extraction from rivers, wells and deep boreholes (Fig. 3.8).

Figure 3.8 Solar pumping system [26].

An inspirational early adopter was Father Bernard Verspieren, who established Mali Aqua Viva, when he found that people were dying because of the failures of diesel-powered pumps [13]. Mali Aqua Viva went on to lead the country's largest water well drilling program and was the beneficiary of the solar panels funded by the Cannes auction (see page 88).

France's Pompes Guinard was a prominent supplier – to Fr Bernard among others (see Fig. 11.10), while the specialist supplier Grundfos developed a successful range of submersible solar pumps.

3.3.2 Health Clinics

The absence of electricity is a particular problem for medical facilities in off-grid rural communities.

Clinics need to maintain vaccines below 4 °C, so another early development application for the PV community was solar-powered vaccine fridges. These were typically small units of no more than 30 L capacity powered by a couple of solar panels and a battery.

Refrigeration was needed not only in the clinics but also for the so-called "Cold Chain" from the vaccine supply center (Fig. 3.9). Transportable solar fridges were

Figure 3.9 Solar-powered Cold Chain vaccine refrigerator by NAPS [25].

developed by NAPS^{qv}, Arco Solar^{qv}, Dulas, and others for all types of vehicle. Perhaps the most iconic was the camel-mounted system, which NAPS' Tchad distributor nicknamed Tapio in honor of chief executive Tapio Alvesalo^{qv}.

3.3.3 Lighting

Worldwide, some two billion people had no supply of grid electricity during the early PV era – and the figure remains well over one billion today. Many of these used candles or kerosene lamps to light their homes, often leading to accidents and fires.

Solar lighting systems were easy to develop, requiring little more than a solar panel, a battery, and a relatively efficient light. Before the advent of high-efficiency LEDs, halogen and low-voltage fluorescent lamps were typically used.

In some regions, sophisticated distribution chains emerged for solar lighting packages.

Before solar-powered consumer products took over (see the following section), solar powered lighting kits were probably the highest volume application for the solar industry. This was the product first deployed by the Solar Electric Light Fund^{qv} (Fig. 3.10).

3.3.4 Rural Community Power

Toward the end of our time frame, these off-grid community applications started to coalesce with the provision of power systems for general domestic, community, and small business use, and this is further discussed in Section 3.6.

Off-grid applications were not confined to so-called developing countries. Remote cabins and holiday homes in Canada, Scandinavia, the United States, and other parts of the world fitted solar panels to provide electricity for the first time, or to supplement or replace diesel generators. Some customers were more unexpected. Legend has it that marijuana growers in California, Colorado, and

Figure 3.10 Gansu family with SELF^{qv} solar panel to light their home [35].

Oregon welcomed solar power as an independent – and untraceable – source of power in remote areas, and could afford the high prices of the time!

3.4 Consumer Products

As solar power became cheaper and more reliable, it was able to replace primary batteries in portable products. These applications proved particularly attractive for less-expensive thin film solar cells, especially in applications less demanding of weather resistance. This chapter deals primarily with household products, while toys and gadgets are mentioned in Section 3.8.

3.4.1 Solar Calculators

The first high-volume application for solar cells was in solar-powered calculators introduced by Sanyo, Sharp, Texas Instruments, and others from the late 1970s, initially using crystalline cells. From the mid-1980s, these progressively migrated to the use of amorphous silicon cells. Calculators consume very little power, so α-Si of relatively low efficiency proved adequate and its superior low light performance suits this usage.

It is this application, more than any other, that led to thin film cells capturing almost 40% of the total market in the mid-1980s, a peak that has never since been matched.

3.4.2 Watches and Clocks

Small quantities of clocks and watches were powered by solar cells as early as the late 1970s, using rectangular monocrystalline cells from Solarex and others.

This market grew, following the introduction of amorphous silicon solar cells, and this too benefited particularly from their superior performance under indoor and low light conditions. This application was championed not only by Japanese producers but also by Chronar^{qv}, as its name implies.

3.4.3 Products for Boats and Recreational Vehicles

Recreational vehicles and boats typically derive their electricity from the engine, when it is running. This doesn't help when the vehicle is not in use, or when the sailor doesn't want to use the engine (Fig. 3.11).

Solar power became recognized as the perfect source in this instance. Panels were soon offered for trickle charging of marine and RV batteries.

In 1982, Michael Pidgeon invented a solar ventilator for boats and caravans powered by a single 3 in. diameter crystalline solar cell. When Intersolar

Figure 3.11 Solar ventilator for boats and caravans [27].

Group^{qv} acquired this business, it extended the application to cars and trucks. During the 1990s, over a million Autovents – a car ventilator powered by amorphous silicon cells – were sold.

3.4.4 Garden Products

The next wave was products for garden use. They gave householders the opportunity to have electrical devices in the backyard without the difficulty and expense of extending mains wiring outside.

The most common products in this sphere are pathway and patio lights. A bewildering array of solar water features are also available, and many other products have subsequently been developed, from electric fences to mole deterrents (Fig. 3.12).

Figure 3.12 Floating pond fountain using amorphous silicon cells [27].

3.4.5 Battery Chargers

An obvious application for PV is to top up rechargeable batteries used in other products such as torches, mobile phones, laptops, and similar. A substantial variety of such battery chargers emerged by the end of our time frame, and it has multiplied further since.

3.5 Grid-Linked Rooftop and Building-Integrated Systems

Power from solar installations on buildings can be supplied directly to the occupants, displacing electricity they would otherwise buy from the grid. This makes such systems economically viable wherever the cost of solar energy is at or below the traditional retail power price, as further explored in Section 6.4.

It is only natural that early PV pioneers looked beyond off-grid applications and thought about solar power for their own homes. Most of the first solar-powered houses were research projects, such as "Solar One" by the University of Delaware[qv] (see Fig. 11.38) in 1973, and Germany's first fully solar house in 1992 by the Fraunhofer Institute[qv] (Fig. 11.41). In Japan, Yukinori Kuwano[qv] installed solar panels on his own roof also in 1992 (see Fig. 10.3), and architecture professor Sue Roaf designed and built her own solar-powered house in the United Kingdom in 1995 (Fig. 3.13).

One of the earliest commercial solar homes was built in the early 1980s by the developer John Long, an innovator in assembly line homebuilding methods. Then in 1988, his company John F. Long Homes with support from the US Department of Energy[qv] built the first solar subdivision, confusingly also called "Solar One." The 24-home block at Osborne Avenue and 71st St in the Phoenix suburb of Glendale is still powered by a ground-mounted solar array.

The first volume roll-out of building-integrated systems was Germany's "Thousand Roof Programme" [37] between 1990 and 1995, which installed solar generators to contribute to the electricity supply of domestic homes. In

Figure 3.13 UK's first architect-designed solar house [36].

due course other countries went on to promote systems for household applications – most notably in Japan, where the "Sunshine Project" [37] in the late 1990s supported 70,000 household installations. Meanwhile, the success of Germany's initial thousand roof trial led to an expanded scheme, entitled "The 100,000 Roofs programme" initiated in 1999 [38].

The use of PV at the household scale became widespread, with supportive policies in many developed countries, as further described in Chapters 8 and 13.

3.5.1 Commercial, Industrial, and Agricultural Buildings

Building-integrated applications are not restricted to rooftops nor to residential houses. Solar systems were installed on offices and many other industrial and farm buildings during our time frame. Photovoltaic power was still relatively expensive, even compared to the retail electricity price, so additional subvention was needed. Several early projects on commercial buildings were supported under research, pilot, and demonstration programs in Europe, Japan, and the United States.

In due course, Flachglas[qv] and others started to exploit the benefits of PV panels as architectural features. If solar panels were used to replace expensive cladding materials such as marble, they could offer economic benefits even if the value of the electricity produced was modest.

Pioneering architects started to embrace the concept, leading to landmark projects such as the German Bundestag by Sir Norman Foster, Mataró library near Barcelona by Miguel Brullet (Fig. 3.14), and the glass roof of Berlin station by Christoph-Friedrich Lange shown in Fig. 4.10.

Subsequently, especially following introduction of feed-in tariffs in the early twenty-first century, the number of building-integrated PV installations, both residential and commercial, has exploded. The German feed-in law of 1991 (see Section 8.4) and US PURPA public utilities act (see Section 8.3) were important for these grid-connected applications.

Net metering, whereby PV power producers can in effect "run the meter backward" when they are generating more power than they need, is also a good

Figure 3.14 Solar wall at Mataró library [39] has cells spaced to let sunlight through.

Figure 3.15 Inverter and meter for early solar building [40].

incentive, because it values all the energy produced at the retail price, rather than paying a lower price for any power supplied back into the grid. This approach was tried during the early PV era but not widely adopted.

One example is shown in Fig. 3.15, where the inverter rack supplied by Alpha Real[qv] for a building in Switzerland also houses the meter with arrows labeled "vor" and "ruck" for forward and backward running meters, respectively.

3.6 Rural Electrification and Local Minigrids

> Connecting homes to the electricity grid is getting very expensive especially as we get deeper and deeper into the rural areas . . . the Solar Home System is safe, affordable and friendly to the environment in which we live.
>
> *Nelson Mandela, 1999 [41]*

Many early rural applications were designed to power single appliances, such as pumps or refrigerators, as described in Section 3.3. In due course, larger systems were developed to meet all the electrical needs of an off-grid rural community. These systems were therefore alternatives to diesel generator sets – the only other electricity option for off-grid communities at the time.

Some of the first installations were funded pilot and demonstration projects such as this solar power station on Kythnos Island (Fig. 3.16). In time, this approach was recognized as a promising way of supplying electricity to remote communities and islands. National and international aid agencies and funders such as the World Bank[qv] supported solar rural electrification projects in many parts of the developing world.

Figure 3.16 European Pilot Project powering Kythnos Island in Greece [42].

One of the most innovative projects of this type was undertaken by a joint venture between Shell^{qv} and the South African electricity utility Eskom. It was the brainchild of financier (and later ACORE president) Michael Eckhart, who was running the Solarbank initiative looking at innovative funding approaches, with input from Herman Bos and John Bonda^{qv}. The initial funding for the Shell-Eskom Solar Energy Company came from the parent organizations, but was progressively recouped as the user paid for power through a prepayment meter. It installed some 10,000 systems between 1998 and 2002. For a more detailed description of the project, see Eckhart's chapter in the *Solar Power* book.[2]

3.7 Utility-Scale Projects

Economically, this is the most demanding application, because it needs to compete with grid electricity at the wholesale bulk price.

There were those who considered centralized power stations generating at the megawatt level to be the holy grail of the industry, directly replacing traditional generation. Others suggest that this approach is regressive, because photovoltaics, being modular, allows distributed generation closer to the point of use to supplant centralized electricity production. Let's leave the argument aside and just look at large-scale centralized PV systems during the early PV era.

The first megawatt-scale PV system was installed by Arco Solar^{qv} at Lugo in California in 1983 (see Fig. 11.2. The company followed this the next year with a 7 MW project on the Carrisa Plain north of Los Angeles (see Fig. 2.12).

The only other megawatt-scale project undertaken in our time frame was by ENEL^{qv} at Serre Persano in Italy in 1993 (see Fig. 11.8).

Quite a large number of grid-connected projects on the order of hundreds of kilowatts were undertaken during this period, particularly under demonstration

---

2 *Solar Power for the World* [43] edited by Wolfgang Palz^{qv}, to which I will refer several times.

Figure 3.17 Topaz Solar utility-scale plant [45].

programs in Europe, Japan, and the United States (see Chapter 13). We would consider those to be community- or island-scale installations, not utility-scale.

Subsequently, from about 2006 onward, centralized utility-scale projects started to become more widespread, and the largest installations are now at the gigawatt scale (1000 MW). There is a rather good book [44] about these large-scale applications.

By coincidence First Solar's Topaz plant (built in 2014) now includes the area where the Carrisa Plain project once stood – highlighted in white in Fig. 3.17. The capacity of Topaz is nearly 700 MW_P and over ½ GW_{AC}, the size of a small nuclear plant.

3.8 Early, Quirky, and Other Applications

A solar cell is simply a source of electricity; it can power any electrical device. Furthermore, PV can be used in parts of the world, where other electricity sources cannot reach. Add into the recipe the amazing inventiveness of humankind, and it is inevitable that the number of applications for photovoltaics is all but infinite.

Some, but not all, of the mainstream applications have been described above. There are plenty of others that piqued the interest of solar pioneers at the time, and we will look at a few more here. Many people were attracted to the concept of clean, abundant energy from the sun but were frustrated that it was too expensive to use in their own homes. To allow these people to engage with PV technology, the industry came up with all sorts of alternative products.

3.8.1 Educational Kits, Gadgets, and Toys

Educational kits, offering experiments that the hobbyist could undertake using photocells have been available for decades. The kit shown overleaf, produced before 1948 (so predating Bell's patent of the solar cell), uses a photoelectric sensor (Fig. 3.18).

Figure 3.18 Early photoelectricity experiment kit, dating from the 1940s [46].

Less academic enthusiasts had to wait until the 1970s, when all sorts of solar-powered fans, music boxes, executive toys, and gadgets came on the market.

The Californian inventor Bill Lamb designed many of these solar consumer products and was thus instrumental in introducing PV to a wider audience.

3.8.2 Headwear, Clothing, and Accessories

In due course this concept spread to headwear; with golf caps and pith helmets being fitted with solar fans to cool the fevered brow of the wearer. More recently, flexible solar panels are being sewn into clothing and bags to recharge mobile phones and other portable electronic devices.

Finally, inventors have worked on ways of combining solar power with vehicles, and we would be remiss not to look at some of them.

3.8.3 Cars and Car Races

The first recorded solar car appeared in 1958 when 10,640 silicon 1 cm × 2 cm solar cells were mounted above an old Baker electric automobile dating from 1912 (Fig. 3.19).

Figure 3.19 Solar-powered Baker Electric car on Waterloo Bridge, 1960 [47].

In the late 1970s, several inventors built solar-powered bicycles and cars, including Ed Passereni, Alan Freeman, and Masaharu Fujita. These were the forerunners to a series of vehicles of increasing complexity and cost.

In 1984, Joel Davidson and Greg Johanson's Sunrunner set the first acknowledged world speed record of about 40 km/h in California. It used Arco Solar^{qv} panels and custom electronics by Brad O'Mara of Photocomm's sister company Balance of Systems Specialists.

Once there are a few cars around, what is the inevitable next step? Races of course!

The first solar car race, the Tour de Sol, took place in Switzerland in 1985, with 29 competitors on a 368 km course from Kreuzlingen to Geneva. A car built by Mercedes, and solar-powered by Markus Real's^{qv} company Alpha Real, won at an average of 38 km/h. Other races followed in many parts of the world, and some became regular events.

Probably the most demanding and prestigious solar car race is the World Solar Challenge in Australia – 3005 km from Darwin to Adelaide. The race was conceived as a way of promoting electric vehicle development by Hans Tholstrup, who in 1982 had driven 4130 km from Perth to Sydney in a solar car. General Motors' Sunracer won the first of these races in 1987 at an average speed of 67km/h. Bern University's Spirit of Biel won the second race in 1990, the slightly slower average of 65 km/h being attributable to the weather. Honda's multimillion dollar cars won in 1993 and 1996, with cells from SunPower^{qv} and UNSW-SPREE^{qv}, respectively, recording averages of 85 and 90 km/h, respectively. The Australian "home team" Aurora won the 1999 race; and average speeds over 100 km/h have subsequently been achieved. Gallus Cadonau's chapter in the *Solar Power* book [43] gives a fuller account of these early car races.

Despite the popularity of these activities, and the sums of money spent on them, no solar-powered cars entered the commercial market during our time frame or subsequently. A small number of production cars were fitted with solar sunroofs from PV Electric, the joint venture between Siemens^{qv} and Arco Solar^{qv}.

3.8.4 Trains, Boats, and Planes

If you can power a car, you can solarize a small boat. Probably the first solar boat was the Solar Craft 1 made in 1975 by Alan Freeman. Two years later another Brit, Edward Hawthorne, crossed the English Channel in solar boat Collinda (Fig. 3.20).

Many other electric boats have since been powered with solar panels, with MS Tûranor PlanetSolar subsequently becoming the first to circumnavigate the world in 2012.

Solar-powered aeroplanes are more demanding because of the extra power and light weight required, but that didn't deter the PV pioneers. The first solar

Figure 3.20 SB Collinda in busy shipping lanes [27].

plane to cross the English Channel was Paul MacCready's Solar Challenger (now on display at the Smithsonian National Air and Space Museum in Virginia) in 1981. The exploits have become ever more demanding latterly, culminating with the circumnavigation of the world by the Solar Impulse piloted by Bertrand Piccard & André Borschberg in 2016 (see Fig. 11.25).

Apart from toy train sets, there were actually no solar-powered trains during our time frame. The "first solar train" was subsequently announced in 2013 in Hungary.

4

Photovoltaic Research

"I'm sure you petroleum folks understand that solar power will solve all our problems. How much money have we blown on that? This is the hippies' program from the seventies and they're still pushing this stuff."
US Senator Trent Lott in a speech to the
Independent Petroleum Association of America [48]

The world owes much to those "hippies from the seventies" who started the journey from expensive esoteric space technology to today's best energy option. A selection of these pioneers are profiled in Section 10.2, and they are indicated by the reference "qv" in the following text.

In the early days of the terrestrial PV industry, it was not easy to persuade the best scientists to get involved at all. Semiconductor technology was still advancing rapidly, becoming mainstream, and was a magnet for good physicists and chemists. Solar energy was widely seen as a long-shot. Charlie Gay[qv] achieved "zero hires" in the first 6 months trying to build the R&D team for Arco Solar[qv] in the mid-1970s and had eventually to recruit his brother [49].

Fortunately, there were those who could see beyond short-term obstacles and imagine the huge potential of a clean energy future. One by one, talented teams came together in the United States, Europe, Japan, and Australia, and the long march to reliable, affordable PV technology began.

4.1   Research Expertise and Drivers

"I was in an industrial laboratory because academia found me unsuitable"
Benoit Mandelbrot [50]

Research in solar technology was undertaken in both industry and academia, and there were close links between the two.

The Solar Generation: Childhood and Adolescence of Terrestrial Photovoltaics, First Edition. Philip R. Wolfe.
© 2018 by the Institute of Electrical and Electronic Engineers, Inc. Published 2018 by John Wiley & Sons, Inc.

As a broad generality, the manufacturers tended to focus on cost reduction, while other groups, mainly in universities, looked at approaches for improving the conversion efficiency of solar cells. Many of these were in due course taken up by the manufacturers, where improved performance justified the incremental cost.

4.1.1 Primary Drivers for PV Research

We will shortly consider photovoltaic technology and the research activities devoted to improving its performance.

First however, let's review the top-level business parameters, informing the directions that research needed to follow. The key characteristics, which most affected the early PV sector, were volume, cost, and efficiency.

Volume

The only PV market at the time, the space industry, demanded, at most, a few kilowatts of solar cells per annum. The terrestrial market would need megawatts per annum (thousands of times more volume) just for off-grid applications. If the grid-connected market became a reality, as most hoped and a few expected, the volumes would be thousands of times larger again at the scale of gigawatts per annum.

Price

Space cells at that time sold on the order of \$100–\$1000s per watt, partly because of the tightly controlled, energy-intensive steps required to purify silicon from metallurgical to semiconductor grade. Early calculations by the US Department of Energy[qv] and others suggested that a cell price around \$2 per watt would be needed to start to compete against mains grid power. Others had showed that the off-grid market could be viable between \$20 and \$5 per watt, depending on the reliability required and the power source being replaced (diesel generators, kerosene, etc.). This meant that cost reductions of about one order of magnitude were needed immediately, just to get into the market, and a further order of magnitude or more would be necessary if solar power was to become mainstream.

Efficiency

Solar cells at the time had conversion efficiencies of about 10%, and low cost processing might have been expected to reduce this figure, at least in the short term. While the absolute efficiency is not critical for most applications because available site area is seldom a limiting factor, more efficient cells lead to smaller systems and so save on structure and cabling costs.

It was these dynamics that led early PV researchers to focus first on developing devices and processes suited to low cost and high volume and second on increasing solar cell efficiencies.

Scientific readers will find that the rest of this section summarizes only the major technological breakthroughs, and that coverage of other advances is sketchy or nonexistent. Nonscientists could find some of the technological details obscure and may prefer to skip to the next section, via Fig. 12.1, charting progress in solar cell technology and the researchers responsible.

4.1.2 Solar Cells

The heart of a photovoltaic system – and the cost-dominant component – is the solar cell, so we will start with that. There have been a huge number of innovative approaches, and we can follow only the main themes, leaving many interesting byways unexplored. The commercial take-up of different materials is mentioned only in passing here, but is reviewed again in Section 5.2. Similarly the efficiency progress for each theme is covered here only fleetingly, but enumerated again in Section 12.1 in Part II.

Monocrystalline silicon was then standard for space cells and so the default material for the early terrestrial industry to adopt. Cells were made from pure single-crystal wafers, at least 300 µm thick, sawn from round single-crystal ingots. There were several potential pathways to lower cost, and most of these were pursued from an early stage.

The main alternatives were as follows:

- Cheaper processing technology for monocrystalline cells in high volumes
- Production of silicon in sheet form to avoid the losses in cutting ingots into wafers;
- Alternative sources of silicon feedstock, avoiding the multiple energy-intensive purification stages required for crystal ingot growing
- Concentrator systems using small solar cells to convert a large area of incidents sunlight
- Alternatives to silicon, especially thin films, that needed only a few micrometers of active material

This section summarizes the research and development activities adopted to pursue these approaches. The extent to which they were successful in accessing the market and reducing costs are discussed in the next two chapters, while the success in improving efficiency is enumerated later in Section 12.1.

4.2 Crystalline Silicon Wafer Solar Cells

We will start with the first of the above options, and then move on to ways of increasing the efficiency of crystalline solar cells.

4.2.1 Monocrystalline Silicon Solar Cells: Cost Reduction

Researchers looked at ways of improving each of the main crystalline cell process steps.

Wafer Supply

The first quick and easy way to reduce the cost of monocrystalline cells was to find cheaper wafers, the silicon disks from which solar cells are made. Almost all monocrystalline silicon at the time was produced for or by the semiconductor industry. Semiconductor chips need to be completely uniform, and so wafers with minor defects or impurities could not be used.

Solar cells are much more tolerant and could successfully be made on many of these off-grade wafers; so PV cell producers bought the semiconductor industry's rejects. While the solar industry was comparatively small, there were plenty of off-grade wafers to go round. Solarex's Peter Varadi[qv] recalls [51] buying wafers literally by the barrel load.

As time went by, more technological approaches were developed and AstroPower[qv], for example, established a process for cleaning up wafers that the semiconductor industry had processed and then rejected.

Crystal Growth and Slicing

While the early PV industry relied initially on bought-in silicon wafers, it progressively moved toward producing its own source material. Some companies installed their own crystal pullers to produce monocrystalline ingots. Others started working on polycrystalline material, as further described in Section 4.3. This led to the need for in-house wafering, and new opportunities for improvement, also described in the next chapter.

Cell Processing (Doping)

The next stage in the process to be improved was the doping of the wafers – introducing the impurities that turn passive silicon into active semiconductor cells. Each manufacturer found its own way of making improvements to this step, which I don't understand and can't describe.

The most significant advances were those that enhanced the throughput rate, reducing the time – and therefore cost – per unit, and those that improved yield, reducing the number of rejects to be discarded.

Solar Cell Contacts

A fruitful area for development was to improve the metal contacts that collect the current at the front and back of the cells. The front contact, in particular, is critical because it needs to cover as small an area as possible to prevent shading the cell; but the smaller it is, the higher is its resistance and therefore the output is reduced. Every early cell producer evolved its own distinctive grid pattern to address this issue.

Contact costs could also be reduced by replacing silver, the contact material used in space cells, by cheaper materials such as solder and aluminum. Further cost reductions were made by better ways of depositing the contact grid. Spectrolab[qv] had already started work on screen printing for this step; and this was taken forward by Arco Solar[qv] and Solec[qv], among others, and progressively became the norm for the industry.

4.2.2 Improving Crystalline Silicon Cell Efficiency

Many groups, in both academia and industry, were interested not purely in cost reductions, but also in increasing solar cell conversion efficiency.

Several of the following approaches were pioneered by Martin Green's SPREE group at the University of New South Wales[qv], the most successful team at obtaining ever-increasing efficiencies through improvements in solar cell design. It achieved an 18% cell in 1983 by combining double-layer antireflection coatings, fine contact lines, and enhanced voltage through surface oxide passivation. Two years later, it added surface texturing to produce the first 20% efficient cell, its so-called passivated emitter solar cell (PESC).

The team moved on to apply a similar approach to the rear surface in their passivated emitter and rear cell (PERC), and latterly achieved 25% efficiency in 2009 by adding local diffusion at the back of the passivated emitter, rear locally diffused (PERL) cell.

As you can see, UNSW-SPREE are also undisputed kings of the acronym!

See Fig. 4.1 for just one example of the excellent graphics they produced to illustrate each new development.

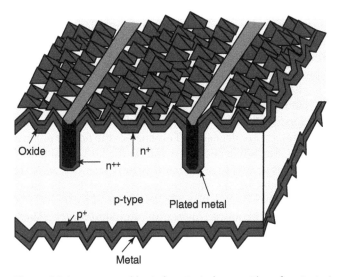

Figure 4.1 Laser-grooved buried contact, shown with surface texturing [52].

Reducing Light Reflection

An early approach to improved efficiency was to reduce the energy lost through light that never entered the cell because it is reflected off the surface. Researchers added antireflective coatings to minimize this problem. Thin layers of transparent materials are deposited on the front surface of the cell and these cause destructive interference for any light reflected at the front and back surfaces of the film.

An additional solution to this issue, adopted by some, was to texture the front surface of the cell, as shown in Fig. 4.1, into a pattern of tiny pyramids – or inverted pyramids – so that any reflected light strikes another part of the cell, instead of heading back out into space.

Improving Cell Characteristics

It was evident from the solar cell I–V characteristic shown in Fig. 2.4 that the power of a solar cell can be increased by increasing the open-circuit voltage and/or short-circuit current and/or by increasing the fill factor to increase the peak power.

The open-circuit voltage is a function of the semiconductor junction. Several approaches were adopted to improve this, for example, more sophisticated processing steps to achieve surface oxide passivation at the front (and later the back) surfaces.

The current produced by a cell is proportional to the illuminated area; so further efforts were devoted to reducing any active area lost, for example, because it is shaded by the front contact grid.

Optimizing the Front Contact

The design of the front contact is a tricky exercise in optimization. The wider the contact fingers, the better they conduct (reducing resistance losses), but the more they cover the active cell (increasing shading losses). Similarly, spacing the fingers more closely reduces the distance the charge carriers have to travel, but increases the shaded proportion. Solar cell producers each reached their own optimum for these dimensions, and the contact materials used.

Another innovation pioneered by the SPREE group is the laser-grooved buried contact illustrated in Fig. 4.1, whereby deep grooves are cut into the cell in which the conductive contacts are deposited. This enables the fingers to have an adequate cross-sectional area, while being very thin in the horizontal plane, thus minimizing their shading.

A different approach to reducing the shading losses caused by the front contact was developed by another leading high-efficiency silicon solar research group led by Dick Swanson[qv] at Stanford University[qv] and SunPower[qv]. Their objective was to develop very high-efficiency solar cells to be used in concentrator systems, where any cost premium could be offset by the fact that much smaller cells would be used. This group developed the interdigitated rear contact, first proposed in 1977 by TRW [53], where both the positive and negative contacts are made on the back of the cell – eliminating any opaque grid on the front.

Using the Albedo Effect

A new approach adopted by Antonio Luque[qv] and others, for locations with high albedo (the reflection of light from the ground or ice and snow), was the bifacial solar cell [54] responsive to light entering the cell from either direction.

4.3 New Approaches to Crystalline Silicon

The research community was not restricted, of course, by making improvements to established monocrystalline silicon technology. It soon started innovating with new approaches, based on both crystalline silicon and other materials.

4.3.1 Faster Crystal Growth

Much of the high cost associated with growing monocrystalline silicon ingots relates to the energy required to keep the silicon feedstock molten over the duration of the crystal growth. Several researchers investigated ways of speeding up the process.

One key protagonist in this approach was Fred Schmid's Crystal Systems, using a process called the heat exchanger method (HEM) originally developed for growing sapphire crystals. The company worked on applying its HEM to silicon, as described in Fred's chapter in the *Solar Power* book [43].

4.3.2 "Solar-Grade" Multicrystalline Silicon Cells

The early terrestrial PV industry had proved, by using rejects from the semiconductor industry, that solar cells are less susceptible to minor crystal defects. This reasoning soon led researchers to develop "solar-grade" silicon, suitable for PV use, but without the expense associated with the high purity needed for semiconductors.

In particular, some showed that the time-consuming process of pulling large monocrystalline ingots could be bypassed by casting silicon in blocks directly from metallurgical grade silicon. Even if this did not result in the whole ingot, being one single crystal it could, if the cooling process was adequately controlled, produce an arrangement of smaller crystals or "grains," adequate for solar cell production. This is variously known as "multicrystalline," "semicrystalline," or "polycrystalline" silicon.

Solarex[qv] started work on this approach under Zim Putney from the late 1970s, setting up a dedicated subsidiary Semix. The chemical feedstock company Wacker[qv] also worked on polycrystalline silicon (Fig. 4.2), and others

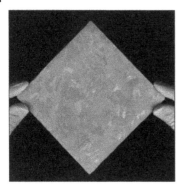

Figure 4.2 Wacker polycrystalline wafer [25]. The mottled pattern is crystal grains.

progressively joined in, including Pragma[qv], through its subsidiary Heliosil, Crystalox[qv] and many subsequently, some initiating other new processes.

Siemens, for example, developed an alternative approach, depositing polycrystalline silicon on thin pure silicon rods using chemical vapor deposition from trichlorosilane (the Siemens process).

Polycrystalline solar cells are typically a few percent less efficient than monocrystalline, because of the higher level of impurities in cast silicon and the impact of the grain boundaries, but in many cases the lower material cost compensates for this difference.

4.3.3 Improved Slicing of Wafers

The next approach to reducing the cost of silicon wafers was to slice them thinner, faster, and with less "kerf loss" – the material discarded from the width of the cut itself.

The introduction of wire saws, initially by HCT Shaping Systems[qv] proved revolutionary, enabling many wafers to be cut simultaneously and allowing the thickness of the wafer (at ~200 µm) and the cut (~140 µm) to be thinner than had been possible using the bladed saws predominantly used at the time. Illustrated in Fig. 11.29, these saws use a thin wire, coated in abrasive diamond powder, wound many times around rollers that maintained the spacing of the slices. Photowatt[qv] and ENE[qv] were the first to use these machines, soon followed by Solarex[qv] and many others.

Meanwhile, Crystal Systems also developed a wire-sawing technique, which it called fixed abrasive slicing technology (FAST).

4.3.4 Ribbon and Sheet Silicon

Another approach to making crystalline silicon wafers more cheaply was to avoid the cost and losses of cutting altogether, by growing the silicon as a thin

Figure 4.3 Octagonal ribbon being grown using ASE's EFG process [25].

ribbon. Like the Czochralski process, this starts with a bath of molten pure silicon, but the ribbon process is designed to draw the material out, not as a "sausage" but as a sheet.

Several alternative approaches have been adopted. Most obvious is the "dendritic web" where two nuclei are introduced into the melt and the crystal is slowly pulled out like a curtain between the two end points.

In 1973, Tyco Laboratories started[1] developing a novel process for growing ribbon silicon in the form of a hollow octagon (seen at the top of Fig. 4.3), which could then be sliced into eight wafers. This was known as edge-defined, film-fed growth (EFG). Their ribbon solar cells typically had a slightly wavy surface (as shown in Fig. 11.36).

Ribbon silicon was promoted by NASA initially for space, for which it ultimately proved unsuitable, because of poor radiation resistance, and then for terrestrial use.

An alternative silicon sheet process was developed by AstroPower[qv] and supported by a program funded by the US Department of Commerce (NIST-ATP). Called Silicon-Film, this process aimed to deposit polycrystalline silicon onto low-cost metallic and ceramic substrates.

Cells based on ribbon silicon gained some penetration into the terrestrial PV market, as outlined in Section 5.2, but have latterly become marginalized, as crystalline wafer technologies have continued to enjoy cost reduction.

---

1 The following year Mobil Oil partnered in this work, forming Mobil Tyco Solar Energy Corporation and the process was later acquired by ASE[qv].

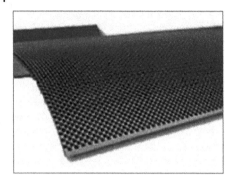

Figure 4.4 Texas Instruments' "spheral" solar cells [25].

4.3.5 Other Crystalline Silicon Technologies

Other even more inventive approaches to low-cost crystalline solar cells have been researched, but none have yet made it into commercial production.

In 1983, Texas Instruments embarked on a very different solar cell technology, which they called "spheral solar" (Fig. 4.4). It used metallurgical grade silicon, formed into spheres of about 1mm diameter through a series of melting and grinding or etching steps. The p-type silicon spheres were then processed by standard diffusion to add an n+ layer, making each sphere an independent PV cell. The cells were then embedded into a thin metal film to provide the interconnections, and making it suitable for flexible configurations.

This development work was supported by the US DOE's PVMaT program. It was subsequently purchased by ATS, soon after they had acquired Photowatt^{qv}, but never transferred into commercial production.

4.4 Other Crystalline Materials

One of the key determinants of the suitability of materials for solar cells is what is known as the bandgap – a measure of the energy, which incoming photons of light require to release charged particles from the material and so a current can flow (remember "The photovoltaic effect" on page 12). Materials with higher bandgaps suit light of longer wavelengths – the blue end of the spectrum – and vice versa.

Crystalline silicon has a bandgap of 1.1 eV, which suits the red end of the spectrum and matches quite well with sunlight as it reaches the earth's surface. Other crystalline elements from Group IV of the periodic table, including diamond (carbon) and germanium, are also suitable for semiconductor production, as is selenium from Group VI. Indeed, the first semiconductor diodes and transistors both were produced on germanium. However, silicon rapidly became the default elemental semiconductor material, in part because of its abundance – the raw material is sand.

Many compound materials (those using more than one element) also exhibit semiconductor and photovoltaic properties. Some are used in crystalline form, while others are produced as thin films (discussed further in Section 4.5).

4.4.1 Gallium Arsenide

So called III–V compounds,[2] such as gallium arsenide (GaAs) and indium phosphide (InP), are good materials for high-efficiency solar cells [55]. Gallium arsenide has a bandgap of about 1.4 eV, matching well with the solar spectrum both in space and on earth. The first gallium–arsenide cells were reported [56] by a team at RCA[qv] involving Dietrich Jenny, Joe Loferski, and Paul Rappaport[qv].

Early processing involved similar steps to monocrystalline silicon: growing and slicing crystals and then doping the wafers. Higher efficiencies were later obtained by epitaxy – growing ordered layers on crystalline substrates.

This complex growth process and high raw material costs make GaAs and other III–V compounds more expensive solar cells. They are also mechanically weaker and more susceptible to imperfections. They are therefore not widely used for terrestrial PV, but have become the material of choice for high-efficiency applications in space, and are used in many concentrator systems. They are also suitable for multijunction devices, as outlined in Section 4.7.

4.5 Thin Film Semiconductor Cells

The prevailing view through most of the early PV era was that crystalline silicon solar cells could never reduce cost to a level to compete for delivering power to the grid. Major PV companies, even those whose commercial business was in crystalline silicon, had substantial thin film research programs. Arguably (and paradoxically), the demise of the leading crystalline producers Arco Solar, and later BP Solar, can be attributed to the failure of their bets on future thin film technology. Japanese companies by contrast survived, thanks to their backing crystalline and thin film approaches more even-handedly. We will return to these dynamics within the industry in Chapters 5 and 7.

Some commentators referred to crystalline silicon as "first generation," expecting that thin films would be second, and heterojunctions would become generation three [57]. With the benefit of hindsight, we know that the PV sector did not develop in such a tidy sequence (even if the research did), so I won't use that terminology as I now move on to discuss thin film semiconductor technologies.

---

2 With one element from Group III and the other from Group V of the periodic table.

4.5.1 Cadmium Sulfide/Copper Sulfide

The early preferred thin film, which was being studied right back in the early 1970s and was seen as a big white hope at the 1973 Cherry Hill Conference (see Section 13.1), combines cadmium sulfide and copper sulfide (Cu_2S/CdS often just known as CdS). These are II–VI and I–VI compounds,[3] with bandgaps of 2.4 and 1.2 eV, respectively.

The first CdS solar cell was reported [58] back in 1954. In our time frame, there were early pioneers in this material on both sides of the Atlantic, notably Karl Boer[qv] in Delaware and Werner Bloss[qv] in Stuttgart. Their work was soon picked up industrially by Shell's SES[qv] and Nukem[qv,] respectively. US-based Photon Power, later acquired by Total[qv], also worked on cadmium sulfide.

Although some product was sold into the market, this material ultimately proved unsuitable for mass production and commercialization, due to "unknown causes of yield, reproducibility and stability problems limited further development," [59], as reported by NASA's Henry Brandhorst or what Nukem called "inherent stability problem of this thin film material due to a structural failure leading to a self-short-circuiting of the solar cell under illumination" [60]. Because use in the mass market of cadmium – highly toxic in its pure form – also had its detractors, this cell material fell out of favor and research stopped.

Cadmium sulfide as a material is still widely used in solar cells, however, featuring as a layer in many cadmium telluride and CIS cells (see both in the following sections).

4.5.2 Cadmium Telluride

Another II–VI compound, cadmium telluride (CdTe) has a bandgap of 1.5 eV and was first reported [61] as a photovoltaic material in 1963. Because it is a direct bandgap material (unlike crystalline silicon), only a few micrometers of CdTe are needed to absorb most of the incident light.

Again, there are several different options for depositing this material, and layers of other materials are added to make it an effective solar cell. We will not go into the processing details here. The toxicity of cadmium and the scarcity of tellurium also meant that CdTe had its detractors.

In the vanguard with this material, in our time frame were the Antec Solar founder, Dieter Bonnet[qv], in academia, and BP Solar[qv] in industry. Though the latter did not pass on its technology, when it shut up shop; others, most notably First Solar, have subsequently developed cadmium telluride technology based on the work of these early pioneers.

---

3 With one element from Group II/I and the other from Group VI of the periodic table.

4.5.3 Amorphous Silicon

After CdS fell from favor, the thin film of choice was amorphous silicon, where the first devices were discovered almost contemporaneously by Walter Spear[qv] at Dundee University[qv], and Dave Carlson[qv] at RCA[qv] in 1975–1976. This gained quick acceptance being made from silicon, an abundantly available material already proven for semiconductor use.

Spear is credited with the first active amorphous silicon device; but it was RCA that applied the material to solar cells and successfully filed patents [62]. The situation was muddied by the fact that the first paper by Spear and his coworker Peter LeComber[qv] was initially rejected for publication (they mistakenly suspected interference from the RCA team). This uncertainty later featured in a series of lawsuits related to RCA's patents, as further described in Section 5.7.

Amorphous silicon (α-Si or aSi) has no crystalline structure, and it was at first a surprise that it exhibits semiconductor properties at all. In fact, the film is an alloy (Si:H) with a proportion of hydrogen, typically produced by deposition from silane gas (SiH_4). Small amounts of other elements are added to the gas during the process to create the required p–n semiconductor junction.

Viable α-Si cells can be produced in films on the order of 1 μm thick, and so require less than 1% of the material used in crystalline cells. Front and rear contacts can be made by depositing conductive layers before and after the amorphous silicon. This sequential layer growth also allows individual cells to be formed by scribing the films. The cells can even be interconnected, without the need for any separate conductive tabs, by overlapping the layers as illustrated in Fig. 4.5, where the bottom contact layer connects through a scribed groove in the α-Si layer to the top contact of the adjoining cell.

Amorphous silicon thus offered the prospect of significantly lower cost, which could more than offset its inferior efficiency – by the late 1980s about half that of crystalline cells.

But characterization work at RCA in 1977 had picked up another issue. Chris Wronski and Dave Staebler noticed that α-Si efficiency declined after exposure

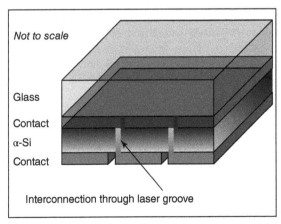

Not to scale

Glass

Contact

α-Si

Contact

Interconnection through laser groove

Figure 4.5 Single-junction α-Si cells showing intrinsic interconnection.

to light. Though Dave Carlson initially discounted this as potential oxidization [63], Staebler and Wronski showed that this light-induced degradation could be reversed by annealing at high temperatures. Their results were validated by others and became recognized as the "Staebler–Wronski effect." This meant that in use, amorphous solar cells progressively declined to around 90% (and in the worst cases, far less) of the initial level at the end of the production line. Wolfgang Palz has a delightfully "un-techy" description [43] of the reason: "The hydrogen atoms in the material are so small and mobile that it is almost impossible to keep them 'quiet' in their positions in the lattice."

On the positive side, α-Si cells were seen to exhibit better performance under low light conditions, as reported by Sanyo's Yukinori Kuwano[qv] [64].

Many companies in the United States, Japan, and then Europe piled onto the amorphous silicon bandwagon, under license to RCA or otherwise. Some used different process techniques, such as the batch plasma deposition of Chronar[qv], while Energy Conversion Devices[qv] (ECD) developed a reel-to-reel process, depositing continuously onto an insulated stainless steel substrate.

Several of the leading researchers also worked on devices with two, three, or more junctions deposited on top of each other. Each junction could be tailored to convert a different part of the light spectrum, thereby improving overall efficiency. Hybrid cells of amorphous silicon with other materials were also developed, notably with CIS by Arco Solar[qv].

4.5.4 Copper–Indium Diselenide (and CIGS)

We will refer to this cell class as CIS, because copper–indium diselenide ($CuInSe_2$) was the first such compound to show photovoltaic properties. However, the introduction of gallium to form $CuInGaSe_2$ (copper–indium–gallium diselenide or CIGS) was soon found to deliver a wider bandgap of 1.3 eV and better material quality; so CIGS has become the preferred combination. For the record, these are I–III–VI compounds, sometimes known as chalcopyrites.

Several production processes have been proven for the CIS absorption layer, including coevaporation of the elements either together or in a multistage process. Additional materials are required to provide contacts and antireflection coatings; so CIS cell structures typically have at least five layers deposited on a glass superstrate.

The first thin film CIS cell was reported [65] by Larry Kazmerski[qv] at the University of Maine in 1976 with an efficiency of about 6%, although crystalline CIS cells had been produced 2 years earlier.

One of the first companies to work on this material was Arco Solar[qv], which was particularly interested in its potential in a multijunction structure with amorphous silicon. There was virtually no commercial exploitation of CIS during the early PV era. Following successive changes in ownership, Arco's technology is latterly being commercialized by the Showa Shell[qv] subsidiary Solar Frontier, and several other producers have also emerged.

4.6 Organic Solar Cells

"The Next Big Thing in Photovoltaics"
How dye-sensitised solar cells were first reported
(. . . and many other "breakthroughs" before and since[4])

You could have made more money than any of the early PV ventures, if given a dollar for every time a revolutionary new solar technology was reported! Many of these "breakthroughs" have never been heard of since. Maybe the journalist had wrong end of stick, or had been bamboozled by an overenthusiastic researcher. Most likely the development never got out of the lab because of efficiency, stability, or cost; or pilot production failed to achieve commercial levels of yield, performance, or economics; or early products didn't deliver realistic levels of output, reliability, stability, lifetime, and so on.

One breakthrough, which has survived, was a new class of solar cell, not based on semiconductors.

In the late 1960s, it had been noted that organic dyes, when illuminated, can generate electricity in electrochemical cells. The University of California at Berkeley studied the phenomenon, and in 1972 demonstrated [66] the principle of power generation via the dye-sensitized solar cell (DSSC). This was a fundamentally different type of cell – more akin to photosynthesis in plants than to semiconductor electronics.

In the following years, more research was done on zinc oxide single crystals, but the efficiency of these devices was poor, because a single layer of dye molecules on a flat surface can absorb only about 1% of the incident light. A breakthrough came in 1991 when Brian O'Regan from Berkeley joined Michael Grätzel[qv] at EPFL[qv] and they introduced nanoporous titanium dioxide electrodes, which increased light harvesting [67], and led to a 7% efficient DSSC. This triggered a boom in research (Fig. 4.6).

Figure 4.6 Dye-sensitized solar cell [26].

---

4 Most recently quantum dot solar cells.

It also created new challenges for the "mainstream" PV industry. DSSC cells respond more slowly than semiconductors, and so needed a new test regime. Furthermore there were no organic reference cells to measure against, and so a new calibration basis was needed. Grätzel credits NREL[qv] in particular with helpfully and patiently working through these issues.

There was no volume manufacturing of organic solar cells during our time frame, although a handful of pilot lines were established, mainly to serve noncritical portable consumer products, such as bags. One or two building-integrated applications also arose – see, for example, Fig. 11.40.

Early production cells had efficiencies of just a few percent, the stability was uncertain, and the packaging required to hermetically seal the cells was unproven, meaning that product lifetime was hard to predict. Subsequently, these issues have continued to be addressed, and laboratory efficiencies have improved markedly for these cells, and derivatives known as "perovskites" – also hailed in their time as "the next big thing."

Uncertainties remain and the jury is out on the eventual market potential for organic solar cells.

4.7 Heterojunction and Multijunction Cells

Heterojunctions are hybrid cells with a single junction incorporating layers of different materials. Devices that stack more than one cell, as mentioned at the end of Section 4.5.3, are known as multijunctions. As has already been noted, most of the compound thin films[5] described in Section 4.5 are strictly heterojunctions, because they incorporate "window" layers of other materials like cadmium sulfide.

Almost all of the solar cells commercialized during our time frame used only one of the materials outlined earlier in this section. However, some initial research was undertaken into heterojunctions. Sanyo[qv] was the first to deposit amorphous thin films on crystalline silicon cells to produce hybrid HIT (heterojunction with intrinsic thin layer) solar cells in 1990. Although these were not widely commercialized until after the end of the century, they have, especially since the acquisition of Sanyo by Panasonic[qv], started to take their place among the more efficient products on the market.

4.7.1 Multijunction Cells

The use of more than one junction can increase the output of a solar cell, because each junction can be tuned to a specific wavelength; so between them, they use more of the incoming light. While the theoretical maximum efficiency for a single junction cell is about 33% (see more in Section 12.2), it has been

---

5 This does not apply therefore to amorphous silicon.

calculated that a cell with a large number of junctions could have a limiting efficiency of 86.8% under highly concentrated sunlight [57].

In addition to multijunctions with two or three stacked amorphous silicon devices mentioned above, some initial research was undertaken into hybrid cell structures, such as the cells incorporating both amorphous silicon and copper–indium diselenide junctions produced, but not commercialized in volume, by Arco Solar^{qv}.

The direction of travel suggests that heterojunctions and multijunctions are likely to form an increasingly important part of the sector as demand for more efficient solar cells develops.

4.8 Solar Modules

Having devoted the majority of this section to solar cells – the core of any PV system – we must recognize that developments of other subsystems were also important to the advance of the sector.

Solar modules are the packages in which cells are housed, so let's deal with them next. The module needs to provide weatherproof encapsulation for the cells, electrical isolation to protect the user, and arrangements for mounting and power offtake – all ideally at minimum cost.

Several interesting approaches were attempted before the industry progressively reached the consensus approach, which makes so many of today's products indistinguishable from each other.

4.8.1 Early Terrestrial Modules

A few companies produced small quantities of solar modules even before our time frame. Modules produced in France by Philips in the early 1960s used 19 mm diameter n-on-p silicon cells of about 6% efficiency with painted silver contacts (see Fig. 3.6). The company progressed to 30 mm cells at about 10% efficiency in 1965 and to 40 mm diameter cells in 1969. The main problems on the earlier models were degradation of the solar cell contacts and cracking of epoxy seals between the plate glass covers and aluminum box frames, later solved by replacing epoxy with rubber.

Perhaps the first terrestrial module in any sort of volume production in our time frame was the Solar Power Corporation^{qv} P1002, comprising five 2 in. round monocrystalline cells wired in series for a nominal 2 V output (Fig. 4.7). The cells were soldered to a fiberglass printed circuit board – the volume production process of that era for electronic devices – and then encapsulated in silicone rubber behind a polycarbonate front cover.

This construction proved adequately robust, which was important because many of the applications at the time were for offshore buoys. However, its

Figure 4.7 An array of seven Solar Power Corp P1002 modules [68] in Lucas' promotion for caravans.

longevity was questionable, because the polycarbonate cover tended to craze when exposed to high levels of UV; and the silicone encapsulant was quite soft (and apparently a delicacy to certain types of rodent!), so was vulnerable if not adequately protected from behind. Bernard McNelis[qv] recalls [68] passing a P1002 round the audience at the 1974 conference in Hamburg[qv]. Delegates were impressed, most having never seen a terrestrial module before, but when it was returned, "one smart guy had gouged out some of the silicone sealant."

Systems designed to charge 12 V batteries used subarrays of seven modules typically mounted on a glass fiber sheet. These could then be used individually to charge batteries on boats and caravans, or mounted to aluminum frames for use in larger professional systems (as shown in the foreground of Fig. 11.22).

The practice of "stringing" cells soon emerged with most manufacturers using conductive tabs to make contacts between the front of one cell and the back of the next. The encapsulants chosen needed to flow freely to avoid distorting the strings – and damaging the fragile cells themselves. The trick was to get any air bubbles out, because they not only looked bad but also shortened module life by facilitating ingress of water vapor. Some producers laid the strings on fiberglass trays, like the Ferranti module in Fig. 4.8.

Figure 4.8 Early 32-cell Ferranti module.

Solar Power Corporation developed a similar construction using a white moulded support tray. The cells were potted in silicone rubber, with a thin layer of harder resin on the front. None of these soft-fronted designs survived long in the market, partly because the surface could not be easily cleaned.

Glass soon became the front surface of choice; it is a cheap, volume-produced material, easily cleaned and widely used in all climates. Europe's first production solar module Philips[qv] RTC's PPX47A used 2 in. (50 mm) round mono-crystalline cells cast in resin between two sheets of glass. Early volume products from Solarex[qv] and Arco Solar[qv] had toughened glass superstrates behind which the cells were encapsulated in silicone, sometimes with a tougher rear film. Suppliers who could offer low iron glass were favored because it absorbs less of the useful light on the way through.

Another feature that module manufacturers soon found they needed to incorporate were blocking diodes and sometimes bypass diodes. Blocking diodes are connected in series with the panel and prevent current flowing the wrong way, for example, batteries discharging back through the panel at night when it is not illuminated.

The need for bypass diodes is more technical and relates to mismatch in the performance of different cells or modules, for example, if one is shaded, while the others are all fully illuminated. In this circumstance, the shaded unit will act as a resistor with current being driven through it by others in the same string. This not only reduces performance but can also damage cells and modules and even cause fires, one of which embarrassingly broke out in the solar array installed in 1982 at the Epcot Center in Florida. Bypass diodes protect modules against this, and some even incorporate diodes within each string to protect individual cells.

4.8.2 Laminated Modules

These technical requirements coupled with the demands of production and cost reduction soon led to design convergence by the leading manufacturers. Glass lamination had become common, particularly within the automotive industry and was soon applied to PV modules. Arco's Bill Yerkes[qv] is credited with pioneering this approach; others soon followed (Fig. 4.9).

Laminated modules typically use a superstrate of toughened glass, behind which the cell strings are held between sheets of vinyl – the preferred options were EVA[6] or PVB[6] – and a rear surface film is applied for weather resistance. The backing material of choice was a proprietary PVF[6] film from DuPont. Most PV manufacturers used white, but Solarex[qv] chose blue (allegedly because you

---

6 Enthusiasts can find the chemical definitions in the Glossary: Chapter C in Part III.

Figure 4.9 Arco Solar's first laminated 1 × 4 ft module [25] (with description for people with great eyesight).

couldn't then so easily see cracked cells); others opted for a transparent rear layer.

The lamination process is undertaken in an autoclave where the assembly is heated at low pressure. The heat causes the vinyl layers to flow, while the reduced pressure sucks out any air from between the layers. Again, it took some time to perfect this process to avoid bubbles and cell damage. Our first lamination attempts at BP caused the cells to crack and split apart disastrously when the vinyl started to flow.

As the designs of different producers converged on this laminated construction – still most widely used today – there were some attempts to standardize on size as well, which would have been convenient for customers. Arco settled on a 4 ft × 1 ft size, but most other module manufacturers used a layout with four rows of nine cells and so were shorter and wider. In the end, no consensus was achieved during our time frame or subsequently.

4.8.3 Double-Glass and Bifacial Modules

Modules where both the front and back surfaces are of glass are preferred in certain applications for architectural and technical reasons. Several producers became expert in this construction.

It is possible to laminate using glass instead of plastic for the rear layer, but it is more difficult because a plastic film can follow the profile of the cells and spaces in between, but glass is flat so requires more lateral flow of the encapsulant vinyl – so more risk of cell breakage.

Producing glass–glass modules for architectural applications, Flachglas^{qv} developed a process based on the production of noise reduction windows. After the solar cell strings were assembled between the glass sheets, a transparent liquid resin was pumped in and then hardened by ultraviolet light. This type of module was used for the iconic glass roof of Berlin's railway station (see Fig. 4.10), designed by architect Christoph-Friedrich Lange, who was awarded the Bonda Prize^{qv} in 2003.

Bifacial solar cells react not only to light falling on the front of the cell but also that from behind. In some applications, the back surface is actively illuminated,

Figure 4.10 PV panels in the roof of Berlin station [69].

using mirrors, for example; in others, it merely capitalizes on the albedo effect – light reflected by snow, ice, and pale rocks and soils.

So solar modules designed to incorporate bifacial cells need a transparent rear surface. Early producers, such as Isofotón[qv], typically used double-glass construction for their modules.

4.8.4 Thin Film Modules

Thin films are deposited as a continuous layer on a solid surface; so the material for this substrate is usually chosen such that it can also act as the front or back of the solar module package.

Glass is a good insulator and has suitable mechanical properties, and it is therefore widely used as superstrate behind which thin film solar cells are deposited. A protective rear surface is then applied to complete the hermetic package. In most cases, this rear coating uses one of the approaches previously described for crystalline silicon modules, such as a resin coating, a plastic film, or lamination with an additional sheet of glass. Finally, the sandwich is framed using aluminum or plastic, as in the Genesis range from Arco Solar[qv] illustrated in Fig. 4.11.

Some thin films were deposited in reel-to-reel processes using plastic and/or stainless steel substrates. Sometimes these films were bonded to a glass front cover, making a rigid module.

Alternatively some manufacturers completed the package with a thin plastic front film. This has the added advantage that the assembly remains flexible, making it suitable for mounting on curved surfaces – such as boat decks – and for certain architectural applications.

Figure 4.11 The 1 ft square amorphous silicon Genesis module.

Figure 4.12 Uni-Solar panels as long flexible strips for roof mounting [25].

Uni-Solar panels from ECD^{qv} consisted of long rectangular strips with wiring at one end, which could be glued to any suitable supporting surface. They were widely used on roofs, where they could be rolled out between raised seems (as shown in Fig. 4.12), and on caravans, motorhomes, and similar applications. Organic solar cells, too are often offered in flexible plastic packages. Flexible modules continue to feature in the market today, but mainly in minority consumer applications.

4.8.5 Concentrator Modules

Concentrator systems focus sunlight from a large area onto a smaller area of active solar cell. Low concentration systems can use standard panels, onto which more light is reflected using mirrors, as illustrated in Fig. 2.12.

But the Holy Grail for concentrators was to achieve high concentration. Dick Swanson described [70] the logic: "Solar cells were two hundred times too expensive. If you can make them five hundred times smaller then, even with added concentrator costs, you're ahead." His early research achieved 28% efficiency at 80× concentration (or "80 suns").

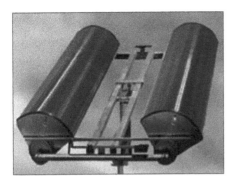

Figure 4.13 Entech's linear concentrator [25].

Even at this level the cells get very hot. So the next challenge for module designers, having achieved uniform illumination at high concentration ratios, is how to cool the solar cells.

Walter Hesse's company Entech decided to extract the heat by passing a coolant through tubes to which the cells were attached, and then using this as a bonus. In July 1982, Entech installed a hybrid PVT (solar electric and thermal) concentrator system at the Dallas-Fort Worth Airport. It produced 27 kW of electric power for the airport and 140 kW thermal in hot water for an airport hotel.

Entech's concentrators used linear lenses, as shown in Fig. 4.13, but most high concentration devices are configured with arrays of square Fresnel lenses, as previously illustrated in Fig. 2.11.

Another way to focus light is luminescent concentration, using layers of fluorescent material, like that illustrated in Fig. 4.14

This approach, not commercialized during our time frame, attaches narrow solar cells round the edge of the luminescent layer. When placed in the sun, this sheet captures most of the light without it passing through or being reflected, so it appears at the edges, where the cells convert it to electricity.

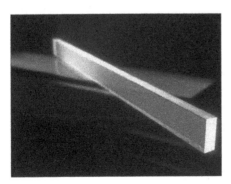

Figure 4.14 Light at the edges of a luminescent concentrator [71].

4.8.6 Custom Modules

Many PV companies developed solar modules especially tailored to individual applications. Navigational aids and offshore platforms, for example, often called for more robust designs resistant to potential collisions, storms, extremes of temperature, and even explosions. Solar modules sandwiched between two thick layers of polycarbonate proved attractive in this market.

Architectural modules typically used the double-glass construction, as mentioned above, but also had to overcome other obstacles – notably a discreet and secure way of bringing the cables out. Standard modules simply incorporated a connection box at the back, where the cables could be connected. This would not suit most architectural installations, so edge connections were developed.

Lightweight portable modules, which could be folded up to fit in a rucksack, were developed for military applications. Modules were also designed for all sorts of leisure applications. For example, flexible units with antislip surfaces are suitable for boat decks.

4.8.7 Module Efficiency

While improvements in solar cell efficiency have been well documented by NREL and others (see Section 12.1), there is sadly much less independent information on the efficiency of solar modules. Of course, this depends primarily on the efficiency of the solar cells inside, but it is also influenced by many other factors.

First, there is lost area in between the cells and around the module frame. Then there are factors like transmissivity of the cover layer and its degradation over time. The rear surface also has an impact; a white film (unlike the blue backing film mentioned above) will reflect some of the wasted light back, which leads to some increase in output.

Even without independent records to prove it, there is support for some manufacturers' claims to have more efficient products. In the early 1980s, the HE (high efficiency) range from Solarex[qv] used monocrystalline cells cut down to squares to maximize packing density. These cells had contact grids with very fine lines to minimize shading of the cells.

Later the Saturn range from BP Solar[qv] used cells incorporating several of the features that had enabled Martin Green's group at the University of New South Wales[qv] to set successive efficiency records. More recently, solar modules based on the high-efficiency rear contact cells from SunPower[qv] and the heterojunction HIT cells from Sanyo/Panasonic[qv] were among the most efficient on the commercial market.

4.9 "Balance-of-System" Components

While the PV majors concentrated their research efforts on the "big ticket items" – the solar cells and modules – as described above, research and development on the balance-of-system elements was not neglected, although it was usually led by others within the supply chain.

4.9.1 Array Structures

The evolution of the solar array mountings was largely a matter of standard structural and mechanical engineering. Fixing a ground-mounted solar array at a given orientation is fairly straightforward. Of course, attention has to be given to wind – and sometimes snow – loading, but the main focus is on ways of achieving the required service life at minimum cost.

Most structures use proven weather-resistant materials; galvanized steel and anodized aluminum with stainless steel fixings. One minor innovation was the introduction of elevated structures to lift the arrays above the worst of the windblown sand and dust (Fig. 4.15).

Most stand-alone PV systems are installed at a fixed tilt angle calculated to give the optimum output. This means that for most of the time, the solar cells are not perpendicular to the sun's rays, and so they do not maximize the area of light harvested. Solar engineers therefore, looked at ways of changing the solar array angle so that it could follow the sun. One early approach to avoid sophisticated mechanisms was an array design that allowed the tilt angle to be changed manually twice or four times per year. The arrays for the European demonstration project on Gavdos Island (see Section 13.3), for example, had a steep winter setting and a shallow summer setting. By changing the array angle around the equinoxes, the arrays can be set to increase the seasonal output.

Continuously tracking structures require more sophistication and a mechanical engineering input. Various methods were devised to track the sun, by both

Figure 4.15 Elevated array structure in North Africa [27].

Figure 4.16 Containerized systems by Solapak[qv] in West Africa [27]. The battery banks and electronic controls are in the containers.

using predefined algorithms and hunting for the orientation where output is maximized.

Great inventiveness was also demanded by some custom designs, such as arrays to be mounted on poles, towers, telephone boxes, offshore oil platforms, and many more obscure requirements.

Companies also evolved various portable and transportable designs. These included containerized systems suitable for rapid onsite installation with the transhipment container continuing to serve not only as a mounting for the arrays but also as a housing for batteries, electronics, and even on occasion standby diesel generators (Fig. 4.16).

4.9.2 Storage Batteries

The battery subsystem was predominantly bought in from established manufacturers and it was they who undertook the lion's share of development work.

Although photovoltaics represented a relatively small part of their market, some battery companies did designate departments or staff to cultivate the sector – notably Chloride, Exide, Gould, SAFT, and Varta – whose Hartwig Willmes was a fixture at PV conferences.

The introduction of sealed lead–acid batteries was a notable innovation of relevance to PV applications, because the maintenance of traditional vented designs increased operating costs and restricted service life. The US-based suppliers Gould National Batteries (GNB) and Germany's Sonnenschein were the first to introduce these in volume for professional PV applications (Fig. 4.17). Brian Harper[qv] at Solar Energy Centre later introduced their own models, particularly in the Middle East.

4.9.3 Electronic Controllers

The design of DC power controllers became a rich seam of innovation, as system integrators learned from field experience how to increase system

Figure 4.17 Rack of GNB's "sealed" lead–acid batteries [25].

reliability and extend battery life. Early controllers were little more than simple battery monitors and voltage regulators to prevent overcharging.

Designs progressively evolved to encompass overdischarge protection, maximum power point tracking, load management, onsite status displays, and interfaces for supervisory control and data acquisition (SCADA).

Alternative presentation of the control equipment also evolved beyond the basic "black boxes" with which the industry began.

Rack-mounted options were developed so that controllers could be integrated into telecommunications equipment racks (Fig. 4.18). Companies progressively offered modular "plug and play" configurations and, for attended installations, more visually user-friendly designs.

A number of suppliers sprang up focusing solely on electronic controls, such as Brad O'Mara's Phoenix-based Balance of Systems Specialists.

Figure 4.18 Rack-mounted PV controller [27].

4.9.4 DC to AC Inverters

Traditional inverters had been designed primarily for continuous operation in other industries. Though quite efficient at full power, these proved relatively inefficient on partial load and so introduced major losses for PV applications, except at noon on sunny days.

Various companies developed inverters better suited to solar applications, with the SunnyBoy from SMA[qv] an early leader for domestic rooftop systems. PV inverters were progressively developed to incorporate maximum power point tracking and to give high-efficiency even at partial load.

4.9.5 Load Equipment

The PV companies' scope of supply normally ends at the power system. Others are responsible for whatever is being powered: the radio, pump, light, or whatever.

However, the achievements of some suppliers in developing equipment expressly for use with PV power supplies should not be forgotten. These include the pumps and vaccine refrigerators mentioned in Section 3.3, and the solar-powered consumer products in Section 3.4.

4.10 Systems Research

This chapter has dealt with research and development on hardware components of PV installations. The final piece of the jigsaw is the design and technology that enables all these elements to work efficiently together as a system.

4.10.1 System Sizing

The calculation of the size of system needed to meet a specific output requirement is critical for stand-alone applications. This is a nontrivial issue given that solar radiation is variable and can't be precisely predicted in advance.

The good news is that on average the radiation data do not vary significantly from year to year, and even the monthly levels are reasonably predictable. Meteorological offices around the world have been collecting solar radiation data for decades. While much of this is now in the public domain, data were scarcer for the first solar generation. Bob Johnson[qv] recalls [34] his euphoria at persuading a contact in Russia to send over a package of local sunshine figures.

At the time, the bible for PV system designers was a publication by the University of Wisconsin [72], giving monthly solar radiation figures for thousands of locations around the world. This covered North America, Europe, and parts of Africa and Asia fairly comprehensively, but was increasingly patchy

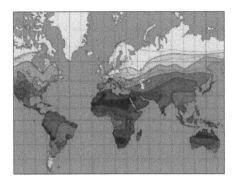

Figure 4.19 World solar radiation map [27]. Darker bands represent higher irradiance in kWh/m^2 per annum.

elsewhere. These and similar figures formed the foundation for the solar radiation maps, which many used for system design (Fig. 4.19).

For professional applications, a more comprehensive approach was needed and systems engineers developed computer programs that used these monthly radiation data to calculate the optimum solar array size, tilt angle and the required battery size to give year-round operation. They could also estimate the probability of power losses.

Unlike grid-connected systems, which aim to maximize annual output and therefore point toward the summer sun, stand-alone systems tend to need a steeper tilt angle facing the winter sun. The design of these computer sizing programs became progressively more sophisticated [73] and enabled designers to optimize the trade-off between cost and reliability.

4.10.2 Reliability, Performance, and Mismatch

There are many factors that can make a poorly designed system very much less than the sum of its parts. Over the years, system developers have identified these, fed them back to the designers of individual subsystems, and adapted their system design procedures to optimize performance. I don't propose to go into detail, but highlight by way of example the issue of mismatch.

When solar cells and modules are connected together in series and parallel, the rule about "the weakest link in the chain" applies. The best cells and modules do not perform to their full potential, but are constrained by the output of the worst ones.

The easiest way of overcoming this is by selection, ensuring that all the cells in a module and all the modules within an array have similar performance characteristics. This doesn't help when one part of a module or an array is shaded, and a hardware solution is needed using a smart wiring configuration, and/or the use of bypass diodes.

5

PV Business and Markets

"Our energy future is choice, not fate. Oil dependence is a problem we need no longer have – and it's cheaper not to.
[It] can be eliminated by proven and attractive technologies that create wealth, enhance choice, and strengthen common security."

Amory Lovins, Rocky Mountain Institute [74]

From a marketing perspective, the early PV industry had a "blank sheet of paper." On one hand, there were virtually no established customers; on the other hand, the prospective market included anyone or anything that needed electricity. It turned out that the challenge was not in finding customers but in persuading them to take solar seriously.

Many people just had trouble believing that a device, silently producing electricity out of thin air without even moving, is not some kind of trick. It sounded too good to be true, what Roberto Vigotti[qv] calls "the Cinderella option." Others worried about what happens at night or when it rains or snows. And that was before even talking about the price.

The early market was heavily reliant on technologically literate customers, who may have read about photovoltaics and were prepared to give it a try. While these early adopters were great friends to the young industry, the most ardent advocates in the long term were often those skeptics who had been persuaded to try one sample system against their better judgment, then discovered to their shock that it actually worked.

5.1 Market Growth

The terrestrial PV market enjoyed progressive growth throughout our time frame (Fig. 5.1). If the start point is low, high growth rates are of course easier to achieve. Nonetheless sustained growth over two-and-a-half decades is

The Solar Generation: Childhood and Adolescence of Terrestrial Photovoltaics, First Edition. Philip R. Wolfe.
© 2018 by the Institute of Electrical and Electronic Engineers, Inc. Published 2018 by John Wiley & Sons, Inc.

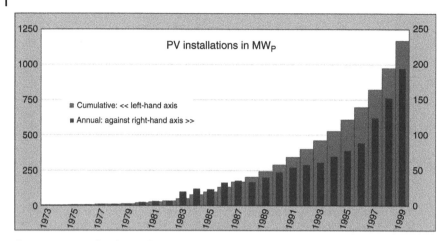

Figure 5.1 Annual and cumulative terrestrial PV installation volume.

noteworthy, and it has continued in most years subsequently, often at even higher rates.

The figures in this section are a fusion collated from various sources, notably PV Energy Systems[qv], with additional input from the International Energy Agency[qv], Strategies Unlimited[qv], Earth Policy Institute, Halcrow, IT Power[qv], Photon Magazine, Sharp[qv], Intersolar Group[qv], Alain Ricaud[qv], and Peter Varadi[qv]. Amalgamating data in this way undoubtedly leads to some inaccuracies, although trends should be fairly representative. Data for the first few years are scarce in any case, and figures in Fig. 5.4 known to be more guesswork than evidence are shown as paler dashed lines.

Many in the industry and beyond made forecasts of the future growth of the sector. These were often written off as "hopelessly overoptimistic," but history has shown that some were quite close, and some proved too conservative, as further explored in Chapter 9.

5.2 Technology Take-Up

Having introduced in Chapter 4 the different classes of PV cell technology, this section will review their respective uptake in the market during our time frame.

When Martin Green[qv] started working on heterojunction cells, he called them "3rd generation," referring to crystalline silicon and thin films as 1st and 2nd generations, respectively. In general, this book does not use that terminology, but for simplicity let's do so here. However, we'll divide crystalline technology into (1a) monocrystalline cells, including ribbon and concentrator

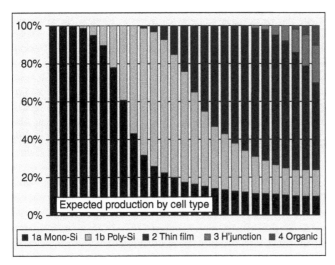

Figure 5.2 Expected PV cell technology penetration trends.

cells, and (1b) polycrystalline cells. We should also add generation 4 for organic cells and other future novel low-cost technologies.

Most observers at the time expected these technology groups to penetrate the market sequentially (hence, 1st generation, etc.). Lower cost polycrystalline silicon, for example, was expected to progressively displace monocrystalline cells, which would then be left with a minority share of the market, addressing only those applications that needed high cell efficiencies. In due course, all crystalline silicon technologies would similarly be displaced by lower cost thin films, and so on. Barring a few particularly prescient experts, the general expectation was that the shares of the global PV market by technology would look something like that shown in Fig. 5.2.

The reality was rather different, with crystalline technologies maintaining the lion's share of the market throughout our time frame, and subsequently. The assumption that crystalline solar cell costs would bottom out somewhere between $1 and $2 per watt – as will be further discussed in Chapter 6 – proved to be wrong. Their costs carried on declining continuously with increasing volume. This meant that the cost advantage of thin film cells didn't actually materialize in practice. Crystalline silicon's market dominance meant that it was being produced in substantially higher volumes and so offset the advantage of the thin films' expected lower material cost.

The resulting breakdown of the terrestrial PV market by technology was as shown in Fig. 5.3. This followed the expected pattern until around 1985, but then monocrystalline silicon refused to concede its market position, and has even latterly remained a dominant contender.

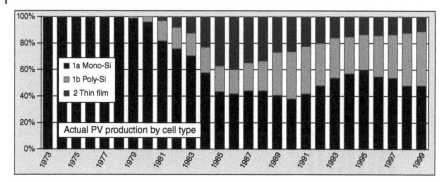

Figure 5.3 Actual PV market penetration by cell technology.

The penetration of thin films into the market show a spike in the mid-1980s, coinciding with the introduction of solar calculators and similar consumer products; and subsequently another peak, albeit at a lower percentage, around 2009, largely due to First Solar's early success in introducing cadmium telluride modules into the European feed-in tariff market. At the time of writing, the market share of thin films in the terrestrial PV sector remains below 10%.

We must not allow the wisdom of hindsight to marginalize the contribution of those who developed new technologies, which in the end achieved low market penetration. Many of those, who took early technology risks, investing tens of millions of dollars in innovative approaches, never got their money back. Yes, the returns would have been massive, if they had achieved patentable low-cost technology. But those who never did still contributed to PV's eventual success.

5.3 Industrial Geography

As already mentioned, a handful of systems had been installed even before 1973. Terrestrial applications in Japan started in the 1960s. A total of 720 photovoltaic systems with a cumulative array output of 13 kW were installed before 1973, of which half were used for lighthouses.

European producer Philips, through its French subsidiary RTC, sold its first commercial systems in 1961. The then USSR is a surprise member of the early adopters list, with installations dating back to the late 1960s, although relatively little activity since. Here a variety of systems, up to 500 W were installed to power inland navigational buoys, irrigation pumps, and communications systems, mainly in the remote arid southeastern region.

Once the terrestrial PV market started in earnest, it became global at a very early stage, unlike most new industries where companies usually start out

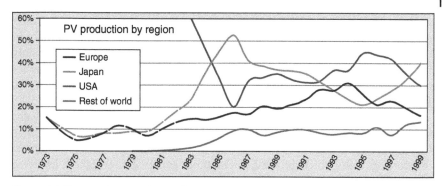

Figure 5.4 Terrestrial PV production by region.

selling in their home market before expanding overseas. This is largely because the research and production was concentrated in the developed world, whereas the most viable early markets were places with high energy costs and a less developed electricity grid.

5.3.1 Cell and Module Production by Region

Let's look first at the geography of early solar cell and module production. This was concentrated primarily in the United States, Europe, and Japan, with the ascendancy swinging between these regions as illustrated in Fig. 5.4.

Note how the first peak for Japan in the mid-1980s coincides with the thin film peak for consumer products shown in Fig. 5.3.

5.3.2 Geographic Markets

There is sadly less independent data about the geographic markets where terrestrial PV systems were installed; thus the following is my estimation of the situation and may prove incomplete.

Through the mid to late 1970s, the early industry was hunting around for applications and addressing markets wherever they could find them. There were early sales in the developed world in the leisure industry, for boats and caravans, and for off-grid buildings and holiday homes.

A progressively larger and more sustainable market that built up in the late 1970s was for professional applications as described in Section 3.2, and this became the major market during the 1980s. These systems too were sold in the developed world, but the largest markets were in developing countries – notably in the Middle East and Africa. The oil-rich countries of the Gulf region proved to be particularly buoyant markets having both the need and the money. African, Asian, and South American markets tended to be more dependent on aid funding.

These professional installations continued to grow in the 1990s with a similar geographic profile. However, that decade also saw significant growth in consumer products, as described in Section 3.4; and these were sold mainly in the leading economies. North America became a major solar product consumer, as did the wealthier countries in the Pacific region, notably Australia, Japan, New Zealand, and Korea.

By the end of the 1990s, residential applications started to achieve significant volumes especially in countries that had dedicated rooftop programs such as Germany, Japan, and the United States. There was good penetration in Switzerland also, which boasted, for a time, some of the highest levels of installed PV per capita.

5.4 Structure of the Industry

The time has come to look at the industrial players in terrestrial photovoltaics. Before going into details on the specific industrial sectors (covered in Chapter 7) and profiling some of the leading companies (Chapter 11), this chapter will look at the types of activity involved.

Figure 5.5 shows the roles of the primary players in the market. Some companies focused on just one role, but many were diversified into several of these activities, as further discussed in the following sections. Let's look briefly at each of these stages in the supply chain.

5.4.1 Silicon Wafer Production

The major usage of silicon wafers throughout the early PV era was outside the solar industry. Major feedstock suppliers and silicon growers were already

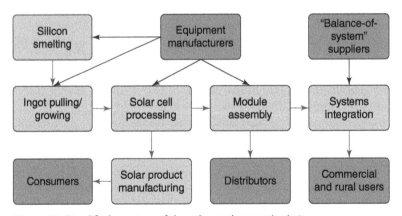

Figure 5.5 Simplified structure of the solar market supply chain.

established. The PV industry itself does not therefore need to incorporate this role; initially solar cell producers bought their wafers from established suppliers in the semiconductor industry.

Before long however, several PV manufacturers elected to bring this task in-house, installing their own crystal pullers and wafer slicing equipment. Others worked on lower cost "solar grade" silicon and installed ingot casting facilities, again with wafering capability. Most of these were integrated with solar cell processing plant, although a number of specialists restricted themselves to feedstock and wafer supply.

5.4.2 Solar Cell Processing

Producing solar cells in what's known as a fabrication plant or "fab," as in the semiconductor industry, is the key process step. Indeed, Jerry Sanders, co-founder of Advanced Micro Devices, famously said in the 1980s, "real men have fabs." Not surprisingly, then, dozens of businesses sprang up to undertake this function in the PV sector, with varying degrees of success.

Cell manufacturers needed to find the optimum trade-off between tight process control to achieve high yield and efficiency and fast processing to deliver high throughput and lower cost. Many cell producers brought in expertise from the semiconductor industry to help develop their wafer fabrication processes. They found the PV sector quite straightforward technologically, but more demanding in other aspects.

Dick Swanson [70] remembers briefing Cypress Semiconductor's experts to help SunPower^{qv} establish a new process line. They asked about the "design rules" – the dimensional accuracy – required. SunPower specified that 100 μm would do. "Don't you mean 100 nanometers?" Cypress answered, accustomed to working to tolerances 1000 times tighter. "How many mask levels? Only 3? We are used to 22 levels; this sounds like kindergarten stuff." The final question: "What daily throughput do you need, then?" One million per day, replied SunPower. "Don't you mean a million per year," said Cypress – not so easy after all!

5.4.3 Thin Film Cell Production

For thin film cell producers, the material production and device processing steps above are effectively combined, because the solar cell junctions are produced as part of the same process that deposits the thin film semiconductor layers.

The various manufacturers between them invented all sorts of different innovative ways of producing films, using feedstocks in gaseous, liquid or even solid form. In most cases, they also needed to develop their own processing equipment, because most were originating completely new technology for which no standard plant was available.

5.4.4   Solar Module Assembly

The next step moves away from semiconductor physics toward mechanical engineering, often with a touch of chemistry still involved. The construction of solar modules, described in Section 4.8, is fundamentally a production line process. It involves interconnecting the solar cells, encapsulating the cell strings behind a suitable cover sheet, and attaching a structural frame and output cable or junction box.

Most solar cell manufacturers also assembled their own panels, but a substantial number of specialist module assemblers also grew up, buying in the solar cells and producing their own panel designs. The main reasons for non-cell producers to establish a module assembly plant were either to increase the level of local or proprietary content, or to make module designs not otherwise available.

Most processes were entirely manual in the early years, but automated production and robots were progressively introduced.

5.4.5   Balance-of-System Suppliers

Several specialist suppliers were founded to provide balance-of-systems equipment specifically for the PV industry, notably electronic controls and mounting structures. Batteries and other standard equipment used in PV systems were usually sourced from established suppliers.

5.4.6   Systems Integrators

Although some leading PV cell and module producers also sold complete systems, many specialist companies emerged to offer engineered PV systems.

Much of the early market was for systems designed to power one particular device: a lighthouse or microwave repeater as described in Section 3.2. An important part of the system integrator's role in this case is to specify the capacity of the solar array and the battery required to meet the load's power consumption.

5.4.7   Solar Product Manufacturers

When lower cost cells made solar-powered products (see Section 3.4) viable, this opened up new manufacturing opportunities. In Japan, it was electronics companies who had first engaged in the PV business, so they were well equipped to produce solar-powered calculators and watches.

Other companies, such as Solar Ventilation in the United Kingdom, designed proprietary solar products. As volumes and competition increased, many solar product businesses subcontracted the manufacturing to the Far East and China, thus establishing themselves as world leaders in low-cost assembly.

5.4.8 Distribution

Most markets, as they mature, develop levels of distribution. You buy your new car from a dealer, not from the manufacturer. So in the PV business, solar companies, initially happy to sell to anyone and everyone, began to establish networks of distributors, especially for their international sales.

Arco Solar[qv], for example, eventually had 96 distributors around the world. It found that in Africa and other emerging economies Singer sewing machine distributors were often the most suitable, because they were "used to selling $200 items in [remote communities]" [75].

Many of the distributors added more value than simply a sales outlet. Some also established themselves as systems integrators. Others added DC lights and other products to the range or introduced packaged systems for the rural electrification market.

5.4.9 Specialization versus Vertical Integration

It is not unusual in the early stages of a new industry for companies to maintain substantial vertical integration to enable them to be active in market development. This was the case for the first solar generation, with the major PV manufacturers making cells and modules and progressively diversifying upward into silicon processing and downward into systems integration.

Similarly, some systems integrators in due course diversified upward into solar module and even cell production, for example Neste[qv], Intersolar Group[qv], and Flachglas[qv].

There were only a few that focused their activities mainly at one point in the supply chain – for example, Spire[qv] in equipment supply and Crystalox[qv] in silicon casting. As industries mature, however, this type of specialization becomes more widespread, as it has subsequently started to transpire in the PV sector too.

5.5 Key Suppliers and Service Providers

Partners beyond the sector are also important for the development of a new industry. We will consider here just a few of those that were crucial to the early solar business.

5.5.1 Materials Supply

The supply of silicon wafers has already been described in the previous section; it depends on the established producers of metallurgical-grade silicon feedstock. The active chemicals for thin film solar cells are typically provided by mainstream gas and chemical suppliers.

Another key material is the glass used for most solar modules, and the role of glass manufacturers is discussed in Section 7.2.

5.5.2 Production Equipment Design and Supply

PV producers received substantial support from established equipment suppliers, particularly in the wafer fabrication and automation sectors. As already mentioned, some also found themselves needing to develop their own equipment.

A few companies decided to exploit the expertise acquired in this way by refocusing their activities on, or spinning out separately, equipment supply businesses. Examples are EPV founded by Zoltan Kiss^{qv} and Crystalox^{qv} (although it later reverted primarily to material supply).

5.5.3 Independent Testing

Manufacturers offer quality assurance to customers, *inter alia*, by having their products independently tested against international standards (developed as outlined in the next chapter).

Initially few organizations were qualified to perform these tests, mainly government laboratories like the EC's Joint Research Centre^{qv} (JRC) at Ispra in Italy, Japan's National Institute of Advanced Industrial Science and Technology (AIST), and NREL^{qv} in the United States.

These centers soon realized that there were likely to be discrepancies between their results, partly because of differences between the light spectrum and the calibrated reference cells each used for its measurements. A round robin to compare standards was organized in 1985 called the Photovoltaic Energy Project (PEP'85). The test laboratories of the European Commission, France, Germany, Italy, Japan, the United Kingdom, and the United States participated. The variances between them were found to be as high as 8%, although most were within 3%. Subsequent rounds, PEP'87 and PEP'93, reduced the variance and involved progressively more agencies.

In due course, other laboratories built up testing capability, including the Fraunhofer Institute^{qv} in Germany, Underwriters' Laboratories in the United States, and, for a time, Royal Aircraft Establishment in the United Kingdom.

5.5.4 Technical Consultancy and Market Information

Experience is scarce in early-stage industries, so many participants need help from consultants to inform them on options and best practice. Some governments set up national solar energy centers; several established consultancy firms started to offer services in the emergent PV field; and a number of new consultancies were founded to focus specifically on the renewable energy sector.

The early entrants had to take advantage of expertise wherever they could find it. On one occasion Solar Energy Research Institute (SERI) of the United States needed a French-speaking expert to assist in a USAID-funded project in Africa. Terry Hart was teaching in Algeria at the time and had written to them requesting technical information. SERI responded asking, "Do you speak French; and do you want a new job?" Hart ended up as their man in Mali, and then joined IT Power^qv, when the project ended.

Companies need market research and other information to support the strategic development business. Several established researchers therefore diversified into the solar sector, like Hal Macomber at US-based Monegon. Meanwhile, a handful of PV specialist information providers also emerged, notably in the United States John Day and Bob Johnson's Strategies Unlimited^qv and Paul Maycock's PV Energy Systems^qv and in Japan Osamu Ikki's RTS Corporation.

In Europe there were no direct equivalents, although EPIA^qv and latterly *Photon Magazine* published market figures. Where governments needed these types of data, they typically turned to their own agencies like NREL^qv and Sandia in the United States, Centre National de la Recherche Scientifique (CNRS) in France, the Energy Technology Support Unit (ETSU) in the United Kingdom, and so on. The European Commission similarly used JRC^qv, but also engaged technical consultancies, including IT Power^qv; Michael Starr, Bill Gillett,[1] and Rod Hacker at Halcrow; Matt Imamura at WIP^qv; and Fred Treble^qv from RAE.

The clients of these information sources sometimes needed to apply secondary interpretation of their own, because these providers approached their topic with differing level of optimism. PV Systems and Strategies Unlimited, for example, were characterized by one, not entirely impartial, observer [76] as "Paul with the cheery stuff, and Bob with the hard stuff."

5.6 Working Together to Advance the Sector

Next we come to the relationships between peers in the sector; and it's a wonder we are not all schizophrenic! On one hand, companies were competing fiercely with each other for their share of a tiny market and researchers were trying to win a technological lead over their contemporaries. On the other hand, knowhow was growing rapidly, so sharing information could save one from falling into potholes others had already found.

The sector soon found a creative tension between competition and collaboration, with strong and lasting fellowships forged between the industrialists and researchers of the day. This chapter looks at some of the settings within which they learnt to work together.

---

1 Bill Gillett went on to lead renewable energy programs at the European Commission.

5.6.1 Conferences

One of the key fora within which solar pioneers met and compared notes were the PV specialist conferences.

The IEEE^{qv} had established a series of US-based solar energy conferences way back in the 1960s, focusing initially on PV for space. The European Commission^{qv} initiated a series of conferences in Europe from 1977. And the Japanese PV conferences, which had been running since 1979, were later broadened to cover the wider Asia-Pacific region. Each of these had a frequency of about 18 months and they were in due course harmonized so that there was a conference somewhere in the world about every 6 months. They also arranged from 1994 to amalgamate together in a single world conference every 4 years. Further details about each of these conference series appear in Chapter 14.

In the early years, these conferences became "must attend" events and the primary rendezvous, not just between the rival companies and universities but also between research and commerce. Attendance soon ran into hundreds, sometimes thousands. This enabled organizers to select attractive venues, usually in sunny parts of the world. They also scheduled plenty of activities outside the lecture hall, including site visits and exhibitions. The social calendar too became an important part of the conference, often starting with a cocktail reception hosted by the local municipality and ending with a full-scale conference dinner.

The opening session at the Athens conference^{qv} (Fig. 5.6) in 1983 was held in the sunshine at the ancient Odeon of Herodes Atticus amphitheater under the Acropolis, hosted by Greece's Minister of Culture, the former actress Melina Mercouri. As the photographs show, the dramatic surroundings made it slightly hard to concentrate on the earnest proceedings on the podium.

The main site visit that year was to Kythnos Island, transported by a new hydrofoil. Unfortunately, high seas prevented it from returning for the pickup. The party was stranded on the hotelless island until a "state of emergency" was declared to allow a ferry to divert to collect them [77].

At the Cannes conference^{qv} 3 years earlier, the organizers had procured several drawings from the renowned cartoonist Jean-Marc Reiser. A grand double auction was conducted by Paul Maycock^{qv} – who proved particularly adept having spent his teenage years deputizing for his auctioneer father – with Len Magid. First, the cartoons were auctioned for cash. Then the money raised was reverse auctioned for solar panels to be donated to Father Bernard Verspieren's Mali Aqua Viva charity (see Section 3.3). The winner – if that is the word – of this second auction was Joseph Lindmayer^{qv}, agreeing to supply Solarex panels at $2.8/W_P$ – way below cost, but matching the US DOE's then future cost target.

Conference organizers were heavily reliant on sizable committees of experts, who put together the conference programs and reviewed the large number of submitted abstracts to select those for publication. For these insiders, the

Figure 5.6 *Al fresco* opening session of the Athens conference [42].

committee dinner was often the ultimate highlight, except for the chairman, who was in danger of ending up fully clothed in the swimming pool.

5.6.2 Joint Research and Information

It is difficult to persuade commercial companies in the same sector to work together, and academic researchers too are reputedly competitive and protective of their intellectual property. Nonetheless, there are benefits to pooling expertise, and many of the government-supported programs required potential competitors to collaborate, as further described in Chapter 8.

The International Energy Agency[qv] in due course provided an additional forum for international collaboration initially in its solar thermal group and later with a dedicated work stream led initially by Roberto Vigotti[qv], further described in Section 8.8.

5.6.3 Developing Standards

Another crucial area for collaboration is in the evolution of the standards against which products should be produced and tested. For the photovoltaics industry, all of these criteria had to be developed from scratch. This gave the opportunity to evolve standards on an international basis, unlike many other sectors where they vary from country to country.

The organization leading development of standards in electrical industries is the International Electrotechnical Commission^{qv} (IEC). It established technical committee TC-82 with delegates from the national standards institutes of interested countries to set standards in the PV sector. The first two important test standards from this group were IEC 61215 for crystalline silicon modules and IEC 61646 for thin film modules. Based initially on the standards used for JPL's Block V procurement (see Section 13.1) and on Qualification Test Procedure 503 from JRC Ispra^{qv}, these specifications deal not only with electrical characterization and performance but also with the structural and environmental conditions that modules must be able to withstand, as outlined in Section 12.3.

As the market started to mature, the industry decided that to make projects more "bankable," quality assurance should extend beyond certifying purely products to the entire generating system. The PV Global Accreditation Program (PV-GAP) was established in 1996 under Peter Varadi, as further described in Section 8.8.

5.6.4 Representative Bodies

Industrial companies in the sector recognized the benefits of working together also to achieve a voice for the sector, particularly in lobbying for appropriate levels of political support. Industry associations were established in many individual countries and across Europe, as further outlined in Section 7.6.

Similar rationale led to the establishment of Eurec Agency^{qv} involving many of Europe's leading renewable energy research establishments.

5.7 Working Separately to Advance the Company

The sector benefited hugely from the collaborations described above, but I wouldn't want to suggest it was all one cosy, happy family. Competition is always ferocious, especially in an emergent industry where many participants are fighting for their very survival.

5.7.1 Competition

The competition between peers doesn't need much elaboration; it's the same in any industry. PV manufacturers fought like cat and dog often cutting prices below cost to grow their market share – not as mad as it sounds in a technology where increased volume is a key factor in reducing cost. And there were similar clashes between systems integrators, distributors, battery suppliers, and so on.

The battles between competing technologies were often just as intense. Crystalline cell suppliers highlighted the instability of amorphous silicon; the use of cadmium was criticized by silicon technology exponents; all semi-conductor technologists disparaged the potential for organic cells; and so on.

Life could be especially tough for systems integrators, often competing with their primary supplier, who controlled the availability and price of the solar modules they used in their systems. As mentioned in Section 5.4, most PV manufacturers had in-house systems engineering – and they were typically much larger companies. One customer was asked why he placed a large order with BP Solar[qv] despite a cheaper and better offer from Solapak[qv]. Answer: "No one ever got fired for buying IBM" – a common truism in the days before personal computing, when International Business Machines had a stranglehold on the global computer industry.

There was also some antagonism between those who put risk capital into manufacturing plant and those who stayed in research as Jim Caldwell recalled [75]. "It used to drive Zoltan [Kiss[qv]] crazy that Stan [Ovshinsky[qv]] kept fleecing the next big company for millions, while he had to actually make and sell things."

Outsiders might be surprised to learn that competition in academia is equally – maybe more – fierce. This is a matter partly of professional pride and partly of funding.

5.7.2 Intellectual Property

The ultimate way to prevent competitors from using your good ideas is to seek patent protection. Many, many patents have been obtained by PV companies and researchers.

Probably the most actively exploited patent was that for a junction within amorphous silicon [62] originally obtained by RCA[qv] and transferred to Solar-ex[qv], when it acquired RCA's thin film research. This was used to prevent Canon from building the amorphous plant in Japan they had planned in collaboration with ECD[qv], and Advanced Photovoltaic Systems from running the one they had built in California – the factory was later bought by BP Solarex. A case against Zoltan Kiss's EPV was settled out of court.

The longest running lawsuit was against Arco Solar[qv], whose CEO got unexpected advance notice sitting in a hotel bar in Tokyo and by chance overhearing lawyers at the next table (maybe there to pursue the Canon case) talking about suing Arco Solar [75].

The suit was compounded by the number of parties involved; BP[qv] had absorbed Amoco, which owned Solarex, which had acquired the patent from RCA. All four companies had retained lawyers, as had Arco, and their partners Siemens[qv] and Showa[qv]. When the judge first entered court, he asked why on earth there were so many lawyers [63]. Long story short – BP Solarex won.

5.8 Corporate and Financial Development

This section shows how the PV market grew in the last century and describes who did what and why they did it. Let's finish up by seeing how things transpired for these industrial participants.

The first thing to say is that it was an era of investment not return. To put it another way, nobody got very rich. The leading PV producers showed financial losses throughout most of the period, even when measured at the operating level before research and capital investment expenditure. In the late 1970s, the annual global market across the entire industry amounted to less than $100 million. The viability of most companies at this time depended on government funding for research and development, demonstration projects, and block purchases.

The emergence of a market for professional off-grid systems (as described in Section 3.2) provided the first significant commercial revenue stream. This afforded the prospect of viability for smaller specialist system integrators but not enough to finance any meaningful solar cell development. It is also an essentially inelastic market; you could drop the price as far as you like, but that would not make any more telecommunications projects happen.

The birth of the consumer product market (see Section 3.4) provided a second commercial revenue stream, typically at higher margins than professional systems. This subsector favored the use of thin film solar cells; so it did little to support developments in the mainstream crystalline cell business.

It was not solely R&D investment that took companies into deficit, some also sold their products at a loss. Overall shortfalls could be quite substantial. When quizzed if the company he worked for had, as reported, lost $60 millions on revenues of $20 millions, Tom Rosenfield replied: "You shouldn't believe everything you hear; figures are often overstated. Our turnover wasn't that high."

BP Solar was the first major PV producer to claim an operating profit in the late 1990s, and other companies in due course followed. It should be emphasized that these calculations would be heavily dependent on the financial treatment of R&D and other investment costs.

The lion's share of the capital the industry needed therefore came from the owners of the businesses themselves; and the majority of that was from the multinationals, as described in Section 7.2. For many of these, there was a point when they decided that "enough was enough," and they then sold, or in some cases closed, the business. Either way, it would be amazing if any recouped more than a fraction of their investment.

Entrepreneurs in the early PV era fared little better, as described in Section 7.1. The main focus of the early solar pioneers, both individual and corporate, was to create a viable solar sector; and most must gain immense satisfaction from the way it has eventually turned out.

I'm not trying to suggest that the whole industry was a bunch of saintly altruists. All were engaged in a high-growth sector; there was good reason to believe that progress would continue, and at some point there would be money to be made. Some were surprised how long this eventually took. Others took reward enough from the allure and excitement of it all – and perhaps from being part of a movement that is so self-evidently a Good Thing.

The combined revenue of companies in the terrestrial PV industry at the end of the century was an estimated $2 billion per annum. This is not a clear-cut number; it depends on how much downstream hardware and services beyond the PV cells and modules are included. The number is tiny in the context of the global energy market, but still shows good progress from a standing start – and the promise of becoming a serious contributor in the future.

5.9 Market Expectation

We finish this section with a review of how developments in the sector affected the perception of the potential eventual contribution of solar technology. On the face of it, the PV market growth during the early PV era, shown in Section 5.1, looks pretty good with increases almost every year over a long period. But that conclusion must be tempered by the low start point; and more importantly we need to know how these actual results compared to expectation.

There was an almost infinite number of market growth projections during this period. You could find impressive "hockey stick" growth curves in every conference, every company presentation. Independent market analysts, too, derived market growth projections, which managers showed to their boards and civil servants to their ministers to justify the next investment request.

For that very reason, it's fair to say that many of these initial projections were hopelessly optimistic. One of the most bullish early scenarios was produced in 1982 by Monegon [78]. This showed exponential growth from a world market of 5 MW in 1981 to 100 GW (1 GW is 1000 MW) by 1996, increasing by over 100% per annum every year for 14 years.

Projections made within the sector itself were rightly treated with caution, while few outsiders believed the industry was important enough to assess at all. It was therefore a red letter day when in the early 1990s one of Shell's series of "future energy scenarios" showed a major future contribution from solar energy (see Fig. 5.7). Yes, this was predicted to occur several decades in the future; but nonetheless here was one of the world's leading fossil energy companies acknowledging a serious future role for solar power. It gave many spectators cause to reassess their stance toward photovoltaics.

Under the Shell scenario [79], solar arrives only about 2020, while the Monegon projection indicated a market of 130 GW by the turn of the century.

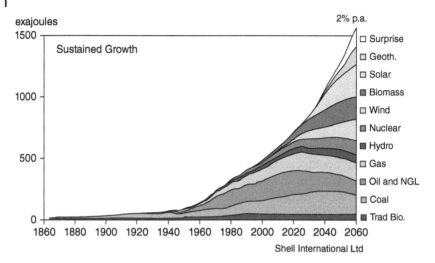

Figure 5.7 Shell future energy scenario [79] from the early 1990s.

Other 1999/2000 forecasts ranged from 200^2 to 10,000 MW$_P$. The actual figure achieved at the end of our time frame in 1999 was almost 200 MW$_P$ – and about 280 MW in 2000.

It is unfair to leave the story at this point, just when the groundwork had been set for more dynamic growth in the future. We will return to this topic in Chapter 9 to see how the projections into the following century turned out.

---

2 Yes, pretty good, this one – the pessimistic "high cost scenario" presented at the 1985 Symposium on Materials and New Processing Technologies [80].

6

Economics of Solar Generation

"Manufacturing costs are too high by a factor of 1000.
Even if collector costs are eventually reduced, photovoltaic devices cannot be considered seriously as a prime source of electricity in the United Kingdom."

UK Government's Energy Technology Support Unit, 1977 [81]

We have seen that PV for terrestrial applications developed from its counterpart in the space sector, where volumes are small and price is a secondary issue. This is why early PV systems were expensive and applications were limited.

"Expensive" is a relative term; we will return to the comparative economics for off-grid and grid-linked applications in Sections 6.3 and 6.4. First, let's look at the factors that drive solar system costs, and see how prices developed after the heady $100+/W level inherited from the space industry.

The measure used here – dollars per watt – is obtained simply by dividing the price[1] in US dollars by the nominal output under standard test conditions in watts peak (W_p) of the solar cell, module, or system. The dollar figures should of course be expressed in real terms, levelized to a given date, since we are considering trends over an extended period and need to eliminate inflation effects.

6.1 Photovoltaic System Costs

This section reviews briefly the cost drivers for each of the major sourcing and production processes previously described in Section 2.5 and Chapter 3.

---

1 Or sometimes the cost, as discussed at the end of section 6.1.2.

The Solar Generation: Childhood and Adolescence of Terrestrial Photovoltaics, First Edition. Philip R. Wolfe.
© 2018 by the Institute of Electrical and Electronic Engineers, Inc. Published 2018 by John Wiley & Sons, Inc.

6.1.1 Silicon Feedstock

In the very early days, when the PV industry was using silicon wafers unsuitable for semiconductor processing, material costs could be negotiated down to a low level – there was no other market for these rejects, and they would otherwise be scrapped. This was fine while photovoltaics used, relative to the semiconductor sector, miniscule numbers of wafers.

As the solar industry grew, it approached the point where there were not enough off-grade wafers to go around; the law of supply and demand pushed prices up. This was one of the drivers that encouraged PV producers to invest in their own silicon growing technology and in the development of solar grade silicon already discussed in Section 4.3.

A single silicon wafer can accommodate thousands of integrated circuits, but it only makes one solar cell; so the PV sector consumes, relative to its revenue, much higher volumes of silicon material. As the industry grew, its offtake of the raw material, metallurgical grade silicon, became more significant. This has led to occasions, albeit after our time frame, when photovoltaics and semiconductors have been in competition at times of silicon shortage – as shown on the right of Fig. 6.1. The semiconductor industry will always win such battles precisely because of the number of devices per wafer and their value, and Siemens Solar[qv] and Sharp[qv] are said [82] to have been among the subsequent losers.

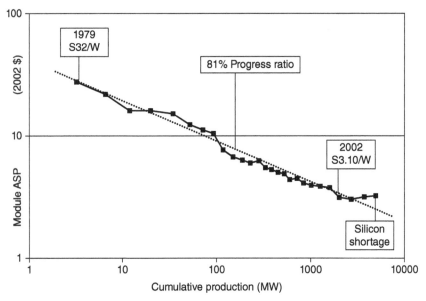

Figure 6.1 Swanson's law – crystalline PV [84] progress ratio is the downward slope of this cost experience line.

6.1.2   Crystalline Silicon Solar Cells

Volume throughput rate is the dominant factor in the cost of processing solar cells, as discussed in Section 4.2. This relationship had already been observed in the electronics industry, where it results from Moore's Law [83], and is expressed in a factor called the "progress ratio."

Experts later postulated that a similar effect would apply to crystalline PV processing, and this has become known as Swanson's Law after a paper [84] by Dick Swanson[qv] in 2006, from which Fig. 6.1 is taken. This shows cumulative volume horizontally and price in 2002 dollars on the vertical scale. Because it is plotted on logarithmic scales, it looks less dramatic than the graph in Fig. 6.2, which uses a linear, or true, scale.

Ignoring minor excursions on either side of the line, this graph shows an 81% progress ratio. In other words, every time cumulative production volume doubles, costs decline to 81% of their previous level. Progress against Swanson's law has been reviewed subsequently [85] and this progress ratio continues.

It should be emphasized that Swanson's law was elaborated only after our time frame. During the early PV era, the widespread assumption was that costs would asymptotically approach a minimum level, below which they could not fall. Paul Maycock[qv] was convinced [86] that this was about $2/W_P$, until persuaded that lower costs were achievable by publication in 1996 of the "MUSIC FM" report [87] for the European Commission[qv].

Another important health warning attaches to these graphs: the figures shown on the vertical axis show price, rather than cost. This would not matter in most cases, where price is above cost by a fairly stable margin, but it will distort the figures in periods when manufacturers sell below cost – as was reportedly the case during the late 1980s, for example.

6.1.3   Thin Film Cells

The cost of producing thin films is also fundamentally linked to volume. Indeed, one would expect that the linkage would be even more pronounced, because the active layers are two orders of magnitude thinner than crystalline cells, making the raw material costs much lower.

It was this factor that fueled the expectation that thin films would progressively increase their penetration by undercutting crystalline cells in the market. As already mentioned, this has yet to happen in practice, because crystalline producers have continued to reap the benefits of increasing volume throughput.

6.1.4   Solar Module Assembly Costs

The costs associated with module assembly are less elastic, being driven mainly by the materials used. Manufacturers have achieved some cost reductions through better design – such as closer spacing of the solar cells – component selection, production throughput, and lower cost materials.

Because solar cell costs have come down faster, the proportion of overall cost attributable to module assembly increased during the period.

6.1.5 Balance-of-System Costs

Again, balance-of-system (BOS) costs, where they have declined, have done very less rapidly than the solar cells; so BOS as a percentage of overall system cost rose progressively too.

This effect was more marked for off-grid systems. Battery costs, in particular, had proved fairly inelastic, because the volumes consumed by the PV market represented a low proportion of total battery sales, so economies of scale had been modest. Reduction in the cost of this subsystem came mainly through improved system sizing techniques, including trading off larger solar arrays against smaller batteries. Recent progress after our time frame, however, has been significant.

For on-grid systems, specialist solar inverters soon became a discrete market sector, as noted in Section 4.9, so the rate of growth of this subsector matched that of the PV market and economies of scale were achieved. Inverter prices, therefore, declined markedly, although not as fast as the solar cells.

Other BOS subsystems achieved some real terms cost reductions as the market grew. Again, this was not as rapid as reducing module cost, so these elements accounted for a progressively higher proportion of the overall system cost, as illustrated in Fig. 6.2.

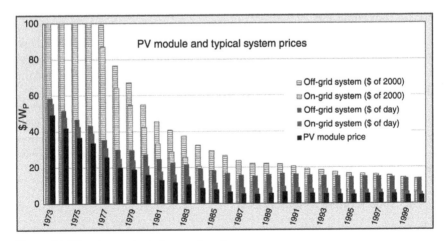

Figure 6.2 PV system costs during our time frame.

6.2 PV System Cost History

To derive the total (uninstalled) system cost simply add the balance-of-system costs to the module, and these are shown in Fig. 6.2. Unlike the progress line shown in Fig. 6.1, these are plotted on a simple linear scale. Costs shown by the solid bars are in "dollars of the day," that is, typical prices paid in the year they appear. Values in real terms, inflation adjusted to 2000 prices, are also shown by the "ghosted" bars behind. The cost reductions achieved in real terms are hugely impressive, and even the unadjusted costs declined almost every year, except when inflation peaked in 1979–1980.

Module costs are shown by the darkest bars. These are the same as those shown on the Swanson's law progress line in Section 6.1, but plotted in dollars of the day to a linear scale.

The solid bars rising above them are typical systems costs (including the modules) shown separately for grid-connected and off-grid systems. The latter are more expensive because of the battery cost. As discussed already, the balance-of-system costs for off-grid systems decline more slowly than modules and on-grid systems. Note that these costs are for "typical" projects; actual system costs vary over a wide range, depending on the location and application.

Shown above these three solid bars are the on- and off-grid system costs, adjusted to 2000 prices. These were so high in the early years they go off the chart. The grid-connected system in 1973, expressed in 2000 prices, amounts to over \$200/W_P. Though horrifically high, it is nonetheless an order of magnitude below the levels reported by the solar cell inventors at Bell Labs. Daryl Chapin calculated that he would need \$1.4 million to power an American home [88], based on costs at \$286/W_P – equivalent to over \$1,800/W_P in 2000 prices.

By the end of our time frame, on-grid system costs had declined in real terms by a factor of over 30 to about \$6/W, thanks to modules now down around \$4.50/W_P. Even more dramatic cost reductions followed in the early twenty-first century.

6.3 Off-Grid Systems: Comparative Costs and Cost Targets

All these dollar per watt figures are very interesting, you're probably thinking, but on their own they are just numbers. The only way of knowing whether it's a good number – whether you're in business – is to compare it to other energy alternatives.

There are established methods to compare the cost of alternative electricity sources feeding the grid, described in the next chapter. We will start, however, with off-grid applications, because these were the ones that first proved viable during the early PV era.

Where there is no electricity grid to rely on, solar systems must be designed to meet the entire energy consumption of whatever equipment they are powering. This can be calculated quite easily (see Section 4.10), if you know the likely sunshine data. A 10 W solar panel will deliver maybe 50 Wh/day at a subtropical site, down to as little as 20 Wh/day in temperate regions. The size of solar system required for any particular load and location is directly proportional to the amount of power required.

The situation is different for most other remote generation options. Primary batteries can supply power for a time, but are economic only for very low-power consumption devices. Diesel generators, by contrast, are available only in certain specific sizes; even if you only want 50 W, you may have to install a 2500 W machine. The operating costs of a diesel genset are also largely independent of its output. Fuel consumption may vary with power delivered, but maintenance costs can be significant whatever the usage.

The cost of extending an electricity grid to a remote site also bears little relation to the energy usage – it depends almost entirely on the distance between the site and the existing grid. An early study in Japan all the way back in the 1960s, when PV costs were astronomically high, showed that solar was a better option than grid extension for power requirements of 4–8 W at sites over 1 km away from grid power.

The net outcome from these dynamics is that the lower the power consumption, and the remoter the site, the greater the likelihood that PV would be cost-effective. Indeed, it is quite easy to calculate, based on power requirement and location, when PV is the most viable option. Early applications, especially in the professional systems market, were based on this calculation.

Adopting this rationale, some crude assumptions, and the cost progression shown in the previous chapter, solar systems for remote off-grid applications were viable at power consumption up to about 1 kWh/day[2] in the late 1970s, several kWh/day in the early 1980s, and up to 100 kWh/day by the early 1990s. By the end of the millennium, solar power was, in many locations, a better option than diesel at any scale.

6.4 Grid-Connected Systems: LCOE and Grid Parity

Assessing the viability of electricity sources, which feed power into the grid, requires a slightly different approach based on calculating the long-term average cost per kilowatt hour.

In the case of a fossil fuel plant, most of this cost comes from the fuel itself and the ongoing operating and environmental costs. Nuclear plants have little fuel

---

2 That is, enough to power a load of about 40 W, if it runs 24 h/day.

cost, but high capital and operating expenditure. For a PV plant, it is different again. The fuel is free and the operating costs are low, but the initial capital expenditure and associated financing costs are higher.

6.4.1 Levelized Cost of Energy

To derive a fair comparison, we calculate a parameter called the levelized cost of energy (LCOE). Take an oversimplified example of a 1 MW plant, costing $1 million, with a lifetime of 20 years, and producing 1500 MWh per year – as it would in, say, Italy or Ontario. If we allow $100,000 per annum for financing and operating expenses, and ignore the discount rate, the levelized cost of energy would be

$$\text{LCOE} = \frac{\$1,000,000 + \$100,000 \times 20}{1500 \times 20} = \$100/\text{MWh} = ¢10/\text{kWh}$$

A more scientific formula is given in the Glossary of terms in Section D of Part III.

6.4.2 Grid Parity

To obtain a target for the required cost of PV, all you need to do is define the electricity price you have to compete with, and run this calculation backward to find the acceptable capital cost. (Okay! You should also add in more realistic operating cost and discount rate assumptions.)

When defining the competitive kilowatt-hour price, it is important to consider who the customer is. If a rooftop PV system supplies electricity to the building's occupants, it displaces electricity they would otherwise buy from the grid; so its value is the retail price of electricity. When the LCOE for solar power falls below this level, it is known as "retail grid parity" [44]. A problem for solar homeowners has been the value of any excess power that they export to the grid. This is often reimbursed by the network operators at a much lower price. One way of overcoming this is "net metering," pioneered by Markus Real[qv] and others, whereby any power exported is netted off the householder's purchased power – effectively (and sometimes actually – see Fig. 3.15) running the meter backward.

In those cases where a generator is installed to feed power directly into the grid – from a free-field solar plant, for example – it displaces electricity from other utility generators. In this case the value is at the wholesale price, typically much lower than the retail price, of course. When solar LCOE falls below this bulk electricity price, it has reached "wholesale grid parity" [44].

The simplified calculation above shows that a capital cost of $1/$W_P$, probably less, is needed to match utility electricity prices on the order of cents per kilowatt hour. Accurate targets were defined in the range $0.70–0.40/$W_P$ as mentioned in the next chapter.

It is clear from Fig. 6.2 that these levels were not achieved during our time frame. By the end of the period however, it was apparent that lower figures could realistically be met in the future. The MUSIC FM study [87] conducted for the European Commission in 1996 identified how solar module costs equivalent to about $0.70/W could be achieved. With the benefit of hindsight, and thanks to Swanson's law, even lower prices have been attained in practice.

6.5 Leveling the Economic Playing Field

The major commercial markets for the first solar generation were in off-grid applications, and we have seen how their scope increased as prices fell. But the world wanted more from the PV sector. To make a meaningful contribution to the replacement of fossil fuels, it has to be able to supply bulk electricity to the grid.

It had been shown that this was economically possible, if production volumes were high enough but, relying solely on the commercial growth of the sector, it would take decades to achieve. Financial support would be needed to accelerate the transition.

In practice, most sources of energy are subsidized. Some politicians complain loudly about the cost of supporting renewable energy but ignore far higher sums spent subsidizing oil and gas exploration and production. Huge amounts are committed, at taxpayers' expense, to nuclear decommissioning and long-term waste storage. It is reasonable that the solar sector should receive funding to "level the playing field," especially, if the subsidy is an interim measure only until grid parity is reached.

Fortunately, there were enlightened politicians in several administrations, who endorsed this viewpoint; so various support mechanisms have been adopted. The main approaches are outlined briefly later, while the individual policy measures introduced in different parts of the world are discussed in Chapter 8.

6.5.1 Cost Targets

The early political support for terrestrial photovoltaics was predicated on the possibility of its eventual contribution to grid power fed into the mains network. As shown already, this implied solar module costs on the order of $1/W. With actual costs starting well over $10/W, supporters needed to see a pathway from here to there.

The United States led in setting targets for cost at both the module and system level. At the European Solar Conference^{qv} in 1977, Len Magid of US DOE^{qv} outlined an aspiration to reduce PV system costs to between $1 and $2/W [89]. The department proposed an interim solar module cost target of $2.8/W_P$ by

1982. Later targets were set with costs declining to $0.70 (in 1980) and eventually $0.15–0.40/$W_P$ [90]. At the same time, one thin film producer, probably with more bravado than substance, predicted he would reach $0.50/W by 1985 [91]. Section 13.1 describes the US DOE program more fully.

While all the target levels have now been achieved, none was reached by the date prescribed. In 1980 equivalent prices, the industry hit the $2.80 level in about 1996, and subsequently achieved $0.70 in 2010 and $0.40 by 2012–2013.

Studies also looked at the size of off-grid markets that might be available at different system costs, and at the costs achievable at those volumes. If the match had been good, the industry could have grown profitably simply from commercial sales in viable applications. It became apparent, however, that development would be slow if funded solely by commercial sales in the off-grid market. That is why the incentive programs described in Chapter 8 were designed to stimulate the additional volume throughput, which has latterly proved so crucial to the success of the industry.

7

Solar Industry Participants

"How to make a small fortune in the photovoltaics business . . .
. . . start with a large fortune."
Aphorism found in many competitive emergent industries

The evolution of the terrestrial photovoltaics industry has been fascinating, with many similarities to the early days of computing and other technology sectors.

In this chapter we will look at the types of industrial players active in the early terrestrial solar power market. Later on, in Sections 11.2 and 11.3 in Part II, we will find profiles of a few dozen of the companies most influential in our time frame

Although the involvement of larger companies soon shaped the development of the sector, much of the pioneering work was by small independent companies – as in most emerging industries – so I shall start with them.

7.1 Where They Came From: Entrepreneurs and Start-Ups

The problem with the French
is that they don't have a word for entrepreneur.
Attributed to President George W. Bush [92]

The role of the entrepreneur is central to the concept of the "American Dream," even if (to the surprise of one former US president) the word originated elsewhere. So it is no shock that many of the early solar entrepreneurs came from the United States.

Most of these first pioneers were already working in PV development in the space sector, recognized the opportunity presented by the first oil shock, and established new companies to address the terrestrial market. Thus, Bill Yerkes[qv] founded Solar Technology International, Joseph Lindmayer[qv] and Peter Varadi[qv] established Solarex, and Ishaq Shahryar[qv] set up Solec all in the mid-1970s.

The Solar Generation: Childhood and Adolescence of Terrestrial Photovoltaics, First Edition. Philip R. Wolfe.
© 2018 by the Institute of Electrical and Electronic Engineers, Inc. Published 2018 by John Wiley & Sons, Inc.

Even earlier, before the first oil crisis, Elliot Berman[qv] had decided that photovoltaics was the field for him and founded Solar Power Corporation.

Outside the United States, Guy Smekens[qv] set up ENE at about the same time, but other pioneers in Europe and Japan chose to operate within the shelter of a larger company. Innovators Claude Remy[qv] and later Tapio Alvesalo[qv], for example, were given, or took, sufficient independence to run their subsidiaries of larger companies as entrepreneurs.

A third group comprised university academics who spun out the fruits of their research into independent companies, as did Allen Barnett[qv] and Dick Swanson[qv] in the United States, and later Martin Green[qv] in Australia. Latterly, there have been others in all of the above categories such as Frank Asbach, Jeremy Leggett, and Georg Salvamoser, but their major achievements fall after our time frame.

Successful companies need both vision and pragmatism. Only exceptional managers can offer both. So the best start-ups had a Yin and Yan double act to provide the required inspiration and acumen. At Arco Solar's heyday, Bill Yerkes could turn to Charlie Gay for a reality check. Peter Varadi, Chuck Wrigley, and Dave Carlson [42] have variously been credited as Joseph Lindmayer's touchstone. The list goes on.

You can't remain a "start-up" for ever. If the sector grows and becomes mainstream – as PV now has – companies have to progress one way or another. There are four probable outcomes: They get bought by someone larger; they grow into a big company, often by merging with others; they find a niche as an independent company; or they die – outcomes described in Sections 7.2–7.5, respectively.

Recent industrial history, particularly in the software and internet sectors, has produced a host of entrepreneurs who have risen to fame and fortune. I would argue that this is very much harder to do in a research-intensive product-based industry, where high growth demands huge amounts of capital for R&D, volume manufacturing, and distribution. Worse than that, the more successfully they grow, the higher the working capital requirement. It is unsurprising that so few entrepreneurs in the PV sector managed to remain independent for long.

Many of the independents that accepted investment from larger companies did so out of dire necessity – in several cases "just in time to meet the month's payroll." Few were paid a generous price. Amoco's initial investment in Solarex[qv] allowed Joseph Lindmayer[qv] to move to a larger house in Potomac and employ a butler, but their later investment rounds were less generous. In personal wealth terms, the early PV industry produced no Bill Gates or Richard Branson.

7.2 Multinational Companies

Major research resources were needed to make PV viable for terrestrial use and to build volume manufacturing plant. All of this is highly capital intensive, so development of the industry would have been much slower without the deep pockets of those multinationals who were brave enough to invest in the early PV era. Their role in the emergent PV market has been one of the most fascinating aspects, and could easily fill a whole book on its own.

Solar modules are glass-fronted semiconductor devices for energy production, so it is no surprise that the participating multinationals came first from the energy, electronics, and glass industries, together with a handful from the aerospace sector, where PV had first been developed. In general – and there are exceptions – the oil companies entered the sector through acquisition; most of the others started with in-house research.

First, since the early solar power business was a cottage industry producing inefficient products expensively and with no clear market proposition, why would a multinational company be interested at all? The answer should be that it has a business plan for diversification, complementary to its core business, with the prospect of substantial profits in the future.

But it seems clear that many were drawn in, at least in part, by other motivations. There were those that saw solar power as a "sexy" emerging sector, where involvement would enhance their reputation or brand; they used it as a public relations exercise. Some energy multinationals seem to have perceived photovoltaics as a potential threat to their core business, and got involved as a way of tracking (some would even say a way of delaying) its progress. But most multinationals viewed their PV investments as "corporate venture capital," what Exxon's Ben Sykes called "probes."

The extent to which these companies viewed renewables as a somewhat risky departure from their core business is exemplified by the paternal aside from a BP board member when Dipesh Shah[qv] was offered the chance to move from the oil business to renewables: "Have a look at it, but don't jeopardise your career" [93].

I must temper the generalizations above with a health warning: industry sectors and corporations should not be regarded as homogeneous entities. Different companies within a sector may well have business strategies that diverge from their peers. Similarly, individual managers within large companies often have agendas not shared by their colleagues.

7.2.1 Oil and Gas Companies

The fossil fuel multinationals were the biggest investors in the emerging PV industry in the 1970s and beyond. This is no surprise, when you remember that we adopted the first oil crisis as our start point. On one hand, the industry became more conscious that oil is a finite resource, and so it needed to look ahead to future energy options. On the other hand, the oil price hike afforded the industry a huge amount of surplus cash. As Jim Caldwell puts it [75]: "There just wasn't enough geology to reinvest all the extra cash in their core business."

Many targeted what they called 'alternative energy' but in many cases this simply meant synthetic fuels and novel sources of fossil fuels, such as tar sands.

Almost every major oil and gas company invested in the PV sector in one form or another. Most prominent were Agip (see Pragma^{qv}), Amoco (Solar-ex^{qv}), Atlantic Richfield (Arco Solar^{qv}), British Petroleum (BP Solar^{qv}), Exxon (Solar Power Corporation^{qv}), Neste (NAPS^{qv}), Shell^{qv}, and Total^{qv} and their activities are each profiled in Chapter 11. In fact, it's easier to list those that weren't involved; in the developed world Chevron and Texaco were the only oil majors to stand aloof, although the former supported research at the University of Delaware^{qv} and has subsequently dipped another toe in the water.

Even those oil companies who did not actively invest in the sector, supported the early industry by using solar generating systems in their oilfields, pipelines, and offshore installations – as described in Section 3.2.3.

We shouldn't overlook how bold it was of the oil companies to embrace the photovoltaic sector with such commitment. They must have realized that it was a very different proposition from their core business. Or perhaps not; Neste asked Tapio Alvesalo^{qv} to rewrite his initial report on the sector, replacing all references to "megawatt-hours" with "tons of oil equivalent."

Given that the leading PV companies through most of our time frame were owned by oil companies, you may well be curious about what happened to them. The expansionist philosophy, whereby they redefined themselves as energy companies – BP even rebranding itself Beyond Petroleum [94] for a short time – receded in the 1980s and 1990s, as memories of the two oil crises dimmed. Most oil and gas companies reverted to focus on their "core business." Those who had not withdrawn from the PV industry by the end of the century, did so soon afterward with the notable exceptions of ENEL, Total Energie [95], and ENI. The differing approaches, which various multi-nationals adopted, when withdrawing from the solar industry, are discussed later in this section.

So, was the fossil fuel companies' involvement in solar power all a conspiracy? While investing in Solar Power Corporation^{qv}, Exxon published advertising [96] saying "When will solar power become a major source of energy? Possibly in the next century." A whole book [97] has been written to suggest that the oil industry had conspired with government and other big businesses to impede the growth of solar energy.

The prevailing view inside the PV industry, however, is that the oil companies are not guilty. Bob Willis^{qv}, former president of SPC said: "Well, the whole time I was president there, nobody at Exxon ever pulled me aside and said, 'Bob, take it easy here. We're trying to push up the price of oil'."

Terry Jester^{qv} spelled out [98] how extensively the oil industry funded early PV research, and how their big balance sheets backed initially risky long-term product guarantees. Charlie Gay^{qv} agrees [49]: "You don't invest hundreds of millions of dollars in an industry to kill it."

7.2.2 Electricity Utilities

I mention the utility sector at this juncture because it seems to be positioned between the primary energy industry and the electronics sector. You would expect that power companies too would be interested in solar energy particularly as a source of electricity. Surprisingly, however, almost none of the major utilities took any active involvement in photovoltaics until very much later.

Probably, the most enthusiastic power company was Californian regional supplier Sacramento Municipal Utility District (SMUD), which established an active solar program and encouraged projects in their region.

The Italian national electricity company (ENEL)^{qv} at first considered PV to be relevant only for islands and other off-grid applications. In due course it started to address rooftop and utility applications, and latterly set up a subsidiary to develop renewable energy projects around the world.

Otherwise most utilities seemed entirely indifferent to developments in the solar power sector, which they saw as irrelevant to mainstream electricity generation. Others seem to have been positively unhelpful, with employees "whose job it is to put you off," as Peter Aschenbrenner put it [24].

Of course, grid-connected projects, especially at the larger scale, demanded the involvement of the grid company, where the connection was made. The individuals handling the projects were usually highly supportive, no matter how indifferent their organizations might have been.

The larger, megawatt-scale projects did get the utilities' attention and Southern California Edison's president attended the opening of Arco's 1 MW Lugo plant. Pacific Gas and Electric (PG&E) provided the land for the follow-on Carrisa Plain project. Two years later they phoned Arco to see if the land was still needed, and were astonished to hear the project was already operational: "We never build anything that fast."

Kay Firor and Dan Shugar from Carl Weinberg's engineering research department at PG&E did later test the value of PV systems by building a 500 kW plant [99] at its Kerman substation in California in 1993. Although the trial proved beneficial for grid support, PG&E did not pursue the concept further after Weinberg's retirement. Shugar went on to establish Powerlight with Tom Dinwoodie and Firor set up Blue Mountain Energy.

It was not until after the utility-scale PV market had subsequently taken off a decade into the new millennium that other utilities, such as EDF, E.On, Iberdrola, RWE, Scottish & Southern, TEPCO, and other Japanese utilities have started to take an interest.

7.2.3 Electronics and Power Equipment Companies

While it was oil and gas companies who were most prominent solar participants in the United States, it was the electronics companies who dominated in Japan and Europe.

Sharp[qv] had been involved in PV even before our time frame and was joined in the 1970s by Fuji, Nippon Electric, and Sanyo[qv] among others. The Kyoto Ceramic company, Kyocera[qv], also got involved because of synergies with the optical components in its photocopying machines.

In Europe, Philips and its French subsidiary RTC were early participants, as were Siemens[qv] and Ferranti. Several US electronics companies took an interest too, notably Texas Instruments and Motorola.

7.2.4 Aerospace Companies

We have already covered in Chapter 2 the space industry's vital contribution to early PV development, so I'm going to say no more about the roles of Boeing, Centralab, Comsat, Heliotek, Hoffman, and TRW, but now will move on to companies developing solar power specifically for the types of terrestrial application, described in Chapter 3.

The German companies Messerschmitt-Bölkow-Blohm (MBB) and Daimler Benz's subsidiary Deutsche Aerospace (DASA) were both involved in terrestrial PV and merged their activities after DASA[qv] acquired MBB in 1989. Spectrolab[qv], although it stayed predominantly in the space industry, does get further mention in Section 11.2 for its contribution to terrestrial PV.

7.2.5 Glass Companies

Major glass companies kept a close eye on the development of the PV industry, partly as a market for its products and partly because thin film solar cells are in many ways similar to other special coatings that glass producers have developed. Their direct involvement in the PV sector has however been fleeting and marginal.

The UK's Pilkington, patent holder on the float glass process, several times considered establishing a PV activity, but in the end left its German subsidiary Flachglas[qv] to take the lead. It had a broad business in the solar sector embracing concentrating solar power (CSP) applications as well as photovoltaics. Pilkington itself became involved latterly with a minority share in a joint venture with Shell[qv] to produce solar cells alongside the Flachglas factory. Another German glass company, Schott, set up a photovoltaics business after our time frame.

Other leading glass companies restricted themselves to supplying glass to module manufacturers with Asahi, Corning, and Saint-Gobain taking strong positions. The smaller Belgian company Glaverbel also benefited because the composition of its glass gave high light transmissivity – well suited to PV.

7.3 Joint Ventures, Mergers, and Acquisitions

It is obvious that start-ups must evolve, as noted at the end of Section 7.1. Maybe less evidently, the solar divisions of multinationals need an exit route too. If they grow too large within the parent organization, they are spun out. If they don't get large enough they will be sold off or closed. And before reaching either of these outcomes many multinationals change their strategy, decline to provide required development resources, or just get bored.

A typical example was the response of the RWE board when Winfried Hoffmann^{qv} proposed an ambitious €50 million proposal for the next development phase for the PV business [100]. "We fully support your proposal," they told him, "but please find another company to pay for it." He did, leading to the eventual takeover of RWE Solar by Schott.

BP Solar started in a similar way after the chief executive of Lucas' solar subsidiary was told it was to be closed to enable the parent company to focus on its ailing core business. Suspecting that BP was interested in entering the sector, he brokered a joint venture; and Lucas were so keen to partner with BP, they stayed on board. Claude Remy seemed to be facing a similar *cul-de-sac* when he heard his boss's boss tell his boss it was time to "arrêter l'aventure Photowatt."[1]

There were many joint ventures and mergers between multinationals or their solar subsidiaries, as detailed in Chapter 11. Some were undertaken in the best interests of the PV business. Others undoubtedly were primarily in the political interests of the parent organizations.

These partnerships took many forms. Some were mergers of two solar subsidiaries – as for CGE Saft and Philips RTC to create Photowatt^{qv} or Nukem^{qv} and DASA^{qv} to create ASE^{qv}. Some were the partial acquisition by one company of the solar subsidiary of another, such as BP's buy-in to Lucas Energy Systems. Some were regional partnerships, like those in Europe and elsewhere of Solarex^{qv} or Chronar^{qv}. Others were joint ventures to develop specific products or markets, as Arco Solar's ventures with Siemens^{qv} and Showa Shell^{qv}, MBB's with Total^{qv}, or Pilkington's with Shell^{qv}.

Few of these partnerships lasted for many years. In practice, the interests of two large and diverse organizations do not remain aligned for long. And the cultural differences were often even harder to bridge.

Most large companies managed their commercial PV business in separate subsidiary companies, but not all. Some of the electronics companies, notably in Japan, felt that the business model was close enough to their mainstream activity to handle solar as a product line within the larger business. The approach to research was more varied, some integrated PV research with the commercial business, others kept it in their corporate research laboratories.

Vesting the solar business in a subsidiary makes it relatively easier to spin-off, sell, or close; and most of the early companies have gone down at least one of these routes, as further discussed in Section 7.5.

---

1 stop this Photowatt adventure.

7.4 Independents and IPOs

We have noted how expensive it is to develop photovoltaic technology and set up volume manufacturing facilities and the even greater cash flow demands, if the business grows fast. This makes it almost impossible for a high growth PV producer to remain independent. All the leading manufacturers were, therefore, acquired or established by large corporations. Smaller companies were able to remain independent, if they stayed in consultancy, service, or downstream markets and out of solar cell production.

A small number of PV companies managed to remain in private ownership throughout our time frame, keeping in-house the pleasure and stress of meeting the payroll every month, notably ENE^{qv}, Intersolar Group^{qv}, and IT Power^{qv}.

If independent companies need infusions of capital to grow, without selling out to a larger corporation, then they must find other sources of investment. Many, of course, tapped the pockets of friends and family, but their resources are seldom very great. The next source is traditional venture capital, and this was widely used to support mid-size companies in the early PV era.

The ultimate source for independent companies needing larger amounts of investment capital is to "go public" and to raise money on a stock exchange through an initial placement offering (IPO). A handful of early PV companies raised money in this way. Spire Corporation^{qv} listed its shares on the "over-the-counter" market in 1983, and Energy Conversion Devices^{qv} floated through an IPO on the US technology exchange NASDAQ in 1985. Both raised additional capital through further offerings subsequently.

Late in our time frame, SolarWorld listed on the German NeuMarkt exchange and AstroPower^{qv} launched its IPO on NASDAQ. Like many before them, they found the pre-IPO "roadshow" an interesting experience. You get typically 1 h to make your multimillion dollar pitch to investment managers, who often react more by gut feel than knowledge of the specific industry. AstroPower's merchant bank decided that investors just wouldn't believe that grid-connected solar power was viable. The company's directors were instructed not to mention it, even though the company had now expanded into the rooftop market.

Latterly, SunPower Corporation^{qv} listed in New York, SMA^{qv} in Frankfurt, and Crystalox^{qv} in London. A number of the newer PV producers are also quoted, mostly in New York, even if they are Chinese companies.

7.5 Where They Went: Divestment, MBOs, and Closures

Having talked about the ways in which various types of company entered, evolved, and survived in the industry, we should reflect on how those that quit did so. If you no longer want your business, or can't afford to keep it running, your choices are to sell it, give it away, or close it.

We covered ways of achieving an industrial sale of the business under mergers and acquisitions. Another alternative for the sale of a business is to allow its employees to buy it in a so-called management buyout (MBO). Claude Remy[qv], Robert de Franclieu[qv], and colleagues did what amounted to an MBO of Photowatt[qv] from Chronar[qv] – although they had previously left the business. Flabeg was created in the MBO of Flachglas Solartechnik[qv], and NAPS[qv] was latterly sold to its management. Some corporates, however, declined possible MBOs, including Exxon for Solar Power Corporation and Atlantic Richfield for Arco Solar.

The doomsday option is to shutdown the shop – almost always the worst financial option. Employees have to be paid off and the assets will realize "fire sale" prices way below what would be achieved selling the business as a going concern. Nonetheless, several companies in the early PV era were closed because either there were no viable buyers, or the parent company's preference.

The oil companies, in particular, often followed this route. Shell's SES, Exxon's Solar Power Corporation, and BP Solar were all liquidated rather than sold. It is hard to understand why they would do this, if viable divestment options existed, unless you subscribe to the theory that the responsible management feared the possibility that a purchaser would make a better job of running the business than they had.

One way or the other, almost all of the early PV producers were divested or closed. Of those listed in Section 11.2, only ENE, SunPower, the Pragma successor EniPower, and the Japanese companies survive.

7.6 Representative Associations

As we move from industry to politics, the representative associations deserve a mention. They provide a focal point for collaboration between the commercial companies, as previously discussed in Section 5.6, but they also have an important role in influencing regional, national, and international policy.

Alongside the main European and North American associations, EPIA and SEIA, to be profiled in Section 11.5, most active countries also had solar industry associations, including CanSIA in Canada, SER in France, BSW-Solar in Germany, Assosolare in Italy, SESI in India, JPEA in Japan, ASIF and others in Spain, and PV-UK.

The approaches they take in representing the sector vary widely, depending on the individuals involved and the relative support of the governments they are dealing with. Because the European Community was largely supportive and positive toward the development of its embryonic PV industry, EPIA's John

Bonda[qv] adopted a cooperative and affable approach in his dealings with the Commission.

At SEIA[qv], Scott Sklar[qv] couldn't always be so conciliatory, in the face of less than consistent support for the sector. He recalls [101] how his sometimes more confrontational approach in dealings with lawmakers and officials often put him in the firing line.

8

Geopolitics of the Early Solar Sector

"If sunbeams were weapons of war, we would have had solar energy centuries ago."

Sir George Porter, Nobel Prize Winner [102]

The speed at which new disruptive technologies can penetrate different market sectors is heavily dependent on the level of regulation.

Consumer electronics, for example, sell into a free market and have famously gained rapid acceptance, leaving billionaires in their wake. Similarly (apart from the billionaires) in the PV industry, the rate of take-up in the consumer market was faster than that in other sectors.

Heavily regulated industries, such as energy, present a greater challenge, primarily because regulation tends to support the status quo. Rapid change in the energy sector could threaten the historical investment of major established corporations, and they have strong political connections. In the electricity sector, access to market provides a further barrier to entry, because electricity networks are de facto monopolies, again with a high level of regulatory involvement.

For these reasons, countries that wished to encourage the photovoltaic industry, and to enable it to make a contribution to their energy mix, realized that they would need to make active interventions in support of its development.

For convenience in this chapter, I will use words like "country," "national," and "government" to refer not only to nation-states but also to international administrations, such as the European Community (as it was then) and agencies of the United Nations, where they are involved in providing support for the sector.

8.1 Global and National Policy Drivers

We have already discussed in Chapter 1 the primary drivers at the start of our time frame, which stimulated the emergence of the terrestrial PV sector. After the first oil crisis of 1973, the Iranian Revolution of 1979 caused a further surge in oil

The Solar Generation: Childhood and Adolescence of Terrestrial Photovoltaics, First Edition. Philip R. Wolfe.
© 2018 by the Institute of Electrical and Electronic Engineers, Inc. Published 2018 by John Wiley & Sons, Inc.

prices, sometimes called the second oil crisis. The energy security driver then receded somewhat during the 1980s and 1990s, but has latterly been escalating.

The climate change driver, conversely, has advanced, thanks largely to a succession of international agreements, which we will summarize briefly. In 1985, the Vienna Convention was signed to protect the earth's ozone layer, and in 1987, the Montreal Protocol was adopted to control chlorofluorocarbons (CFCs) and other ozone-depleting substances.

The United Nations Conference on Environment and Development (the Earth Summit) in Rio de Janeiro in 1992 was the largest UN conference in history with representatives from 172 countries. It produced the Rio Declaration, which outlined principles for a new global partnership toward sustainable development, and Agenda 21, a specific action plan.

At the third United Nations Framework Convention in 1997 (COP 3), UN member states adopted the Kyoto Protocol on Climate Change, which set a target for reducing greenhouse gas emissions in developed countries. This laid the groundwork for efforts on collaborative emissions reductions leading latterly to the Paris Agreement at COP 21.

8.1.1 National Policy Drivers

National governments and international bodies set policies to respond to these global drivers and to benefit the economies and citizens for which they are responsible.

Policy in the solar power sector is normally directed to one or both of these key objectives: to support national companies and research organizations and to increase the level of national solar generation. Many administrations also have a subsidiary objective to use PV to support overseas development.

The PV industry has enjoyed national support in a multitude of forms. The word "enjoyed" may be inappropriate – living off national subsidies is not beneficial in the long term, if the support does not lead to future unsubsidized commercial applications. Several solar companies in our time frame, and subsequently, have collapsed when national support policies ended or changed.

8.2 Incentive Mechanisms

We will summarize the common approaches that have been used, so that we can then focus on specifics of the main support programs around the world.

8.2.1 Direct Research and Development

Some national research laboratories directly undertake photovoltaic research and development. This may be done in partnership with, or for the benefit of, individual companies in the field.

Alternatively, the fruits of successful research are licensed by the research organization to commercial companies – to my knowledge, never commercialized by the nationalized entities or governments themselves.

8.2.2 Support for Industrial and Academic R&D

Many governments provide support in the form of funding and/or technology for companies, universities, and research organizations in the sector. This is usually directed toward specific programs or projects, but may alternatively be unhypothecated grant funding.

This type of support is most applicable at the early, precommercial phase of a development program, so there is typically no penalty if the project does not deliver the anticipated results.

8.2.3 Support for New Manufacturing Facilities and Scale-Up

Governments often support their companies in establishing new and upgraded manufacturing facilities.

Again, this approach usually has no penalties for failure so, while useful for the beneficiary, it might not incentivize competitiveness and efficiency.

8.2.4 Block Purchases

A good way to support the transition to a commercial industry is to bulk purchase a significant volume of finished products such as solar modules. Underwriting the market in this way enables companies to justify investment in manufacturing and distribution facilities.

The extent to which this approach also acts as a driver for product quality and cost reduction will depend on the terms of the block purchase, and the way in which the solar modules are then used. This approach supports PV device production, whereas the other mechanisms described further apply to complete solar power systems.

8.2.5 Grant-Funded Pilot and Demonstration Projects

Another alternative is to support the establishment of complete solar power projects, including not only the solar modules but also balance-of-systems equipment. This approach combines bulk purchasing with support for systems development. Examples are rooftop programs in Japan and Germany and the US and EC pilot and demonstration programs. This option does place specific demands on the quality of both the solar modules and the overall system.

Most of the pilot and demonstration programs supporting early systems deployment made a direct grant to the project developer meeting a proportion of the capital cost of the system. The early rooftop programs also gave grants, but in that case they were often paid to the system purchaser rather than the supplier.

Grants are appropriate at early stages of an industry, because they provide direct funding and are simple to administer. If there is a disadvantage, it is that they give little incentive for efficient system performance or for cost reduction.

8.2.6 Power Production Subsidies and Feed-In Tariffs

Support in the form of a subsidy for power generated is arguably the most suitable at later stages of development, once technology is proven. Such subsidies are designed to offer a guaranteed revenue per kilowatt-hour of output.

They, therefore, provide an incentive not only to supply equipment at competitive cost but also for efficient system design and long product lifetime. The production subsidy mechanism that developed furthest and started to be used in our time frame was the feed-in tariff (FiTs).

Feed-in tariffs reward system efficiency, as mentioned above, and also provide competition, as the payments are made to the system purchaser. A few local feed-in tariff mechanisms were introduced during the last century, notably in Aachen. National FiTs programs, led by Germany, were implemented only after our time frame. They proved outstandingly successful in growing volume for the industry and thereby reducing costs. Indeed governments soon had to contain the level of subsidy by introducing so-called "degression."

8.2.7 Later Measures

Many incentive mechanisms have subsequently been used, with the focus latterly on reducing the financial burden on taxpayers.

Apart from those mentioned already, the four main types of support are as follows: supplier obligations, requiring electricity suppliers to source a proportion of their energy from renewables; reverse auctions, where suppliers are invited to offer the most competitive electricity prices from new generating plant; fiscal incentives such as tax offsets either for the capital cost of the installation or for the energy it produces; and guarantee schemes, which underwrite the loans required to finance projects.

Apart from these later measures, all of the above approaches, and others, were adopted in various parts of the world to stimulate the growth of the early terrestrial PV sector.

8.3 Policy in the United States

US policymakers were fastest out of the blocks in identifying the benefit of supporting the solar industry in response to the first oil crisis. A conference was convened in

October 1973 at Cherry Hill in New Jersey [59], involving key individuals [103] with knowledge of photovoltaics from the space sector. This identified a 10-year program to make PV suitable for terrestrial use, and set initial targets.[1]

To identify the work streams required, projects were established at government laboratories such as Sandia National Laboratories and NASA's Jet Propulsion Laboratory[qv] (JPL) under John Goldsmith[qv]. One of the transformational initiatives overseen by JPL was the "Block-buys," which started in 1975 and ended in 1978.[1] These not only provided manufacturers with offtake volume for their production plants, but they also established quality requirements, which later formed the basis for the IEC's international standards.

Following the election of President Jimmy Carter, a federal Department of Energy[qv] was established. This took on the coordination of the various programs under Paul Maycock[qv]. The main elements were advanced and long-range research, productionization, testing, and systems engineering and applications.[1] This was designed as a close partnership with industry, where "the Department of Energy stimulates development and use of PV systems; and industry performs most government-sponsored work and manufactures and sells products."

The subsequent evolution of US solar energy policy is really a tale of four presidents, as reflected in the swings of expenditure shown in Fig. 8.2. Here is a selection of the apparent high and low lights.

8.3.1 Jimmy Carter Presidency 1977–1981

> Because we are now running out of gas and oil, we must prepare quickly for a third change . . . to permanent renewable energy sources, like solar power.
>
> *Jimmy Carter; televised speech, April 18, 1977*

The next president to be elected after the first oil crisis and the incumbent during the second one, Jimmy Carter, can take much of the credit for placing the United States in pole position in the early PV era. As far as the need to diversify energy sources goes, Carter "got it." He had solar panels installed on the White House (Fig. 8.1).

President Carter created the Department of Energy under James Schlesinger, and ensured it was adequately funded to undertake the development program outlined above. He also established the Solar Energy Research Institute (SERI, which later evolved into NREL[qv]).

The Carter administration introduced energy tax credits as part of the Energy Tax Act of 1978. This created a tax credit for business investing in energy equipment other than oil or natural gas. For solar and wind energy, these credits

---

1 Further details of these targets and programs are given in Section 13.1.

Figure 8.1 President Jimmy Carter and White House solar thermal panels [25] installed in 1977 (later removed under Ronald Reagan).

could even be received as a payment if the beneficiary had no tax liability to offset. This mechanism subsequently formed the basis of the Investment Tax Credits.

Another regulatory measure enacted under the Carter administration, whose importance has only subsequently become apparent, was the Public Utility Regulatory Policy Act [104] (PURPA) of 1978. To paraphrase, the Act compels utilities to purchase energy produced by renewable generators if they can deliver at a price below the utility's avoided cost.

8.3.2 Ronald Reagan Presidency 1981–1989

"Eighty percent of air pollution comes not from chimneys and auto exhaust pipes but from plants and trees."

Ronald Reagan; radio broadcast, 1979

Carter's successor did not share his sympathy for new clean sources of power. He stripped the White House roof of its solar panels, reduced tax credits, and slashed the US DOE's renewables budget by two-thirds.

Paul Maycock left and his successor, Mort Prince lamented: "I was losing all my best people." Industry too was dismayed. "If Reagan wanted to get out of solar," said Joseph Lindmayer[qv], "at least he could have done it like Vietnam and called it a victory. Instead he pulled out as if solar was a loser."

This was a low period for US national policy toward PV, although in the commercial market it retained a dominant position.

8.3.3 George Bush Presidency 1989–1993

Branded "not as bad as expected" [101], George Bush Senior's presidency took a business approach to most things, including renewables. It increased and stabilized the US DOE's budgets and converted SERI to NREL[qv], now formally a government laboratory, thereby making its status and funding more secure.

The US DOE's support program was transformed into the PV Materials and Technology Programe (PVMaT), under the management of NREL. The Energy Policy Act of 1992 introduced the first Production Tax Credits for renewable energy generation.

8.3.4 Bill Clinton Presidency 1993–2001

In Bill Clinton's administration, the president tended to focus on business, while Vice President Al Gore was concerned about the increasingly evident effects of climate change – a field in which his fame has subsequently increased.

The administration reintroduced more generous tax credits and encouraged US states to do more to support renewables. In 1997, the Million Solar Roofs Initiative was introduced. By the time the final report [105] was published in 2006, the initiative had stimulated 377,000 residential solar installations with $16 million of federal funding.

8.3.5 Other US Government Activities

The US DOE led on development of the PV sector, but was not the only ministry involved. All government departments were encouraged to deploy solar generation in their own facilities and to promote it in their areas of activity. SERI produced a comprehensive report [106] to help them.

The Department of Commerce set up in 1991 the Advanced Technology Program (NIST-ATP) within the National Institute of Standards and Technology to stimulate early-stage advanced technology development that would otherwise not be funded. Photovoltaics was amongst the technologies that it supported.

USAID, the US government's overseas aid agency, has supported much PV work in developing countries. Again, this was assisted by SERI.

8.3.6 Later and Regional Activities

Subsequently, after our time frame, further federal interventions have stimulated the deployment of PV systems in the United States. Fiscal measures such as production tax credits and investment tax credits have been particularly influential.

Not all US policy is determined at the federal level and several states have been particularly prominent latterly in supporting the deployment of solar power plants, particularly through renewable portfolio standards.

8.4 Policy in Europe

For members of the European Community – as what is now the European Union was known then – two levels of policy applied: that of the national governments and the Community-wide initiatives administered by the European Commission in Brussels. We will start with the latter, because many countries viewed Europe as the policy leader for renewable energy development.

Legislative edicts from the European Commission – such as the 2001 and 2009 Renewable Energy Directives – have become strong drivers latterly, but it was other more direct measures that were relevant to the early PV sector.

8.4.1 European Commission

Setting policy at the community level requires consensus between the member states. This can be a protracted process, but has the advantage that once adopted, it is not subject to sequential elections in any one country (as for the United States above), so should be more stable.

Another feature is the EC's penchant for requiring collaboration between partners from different countries within the projects it supports. This provides benefits in knowledge sharing, but is rather an anathema to naturally competitive businesses (and academics) and adds to the administrative burden.

While the early renewable energy sector was in the R&D phase, DG-XII took the lead in supporting its development.

Directorate General for Science Research and Development DG-XII under the leadership of its first Director General, Günter Schuster, had established in the early 1970s its Research and Technology Development Programme (RTD). This started with an environmental program, followed in 1975 by an energy program, led by Albert Strub. Wolfgang Palz[qv] joined 2 years later to lead the solar work. The RTD was undertaken in a series of 5-year programs, the first of which (1975–1979) had a solar budget of €50 million.

DG-XII established a strong renewable energy capability at the Joint Research Centre at Ispra[qv], which inter alia played a leading role in developing standards for PV products and qualification testing on behalf of European producers [107]. The solar program supported a wide variety of science and technology development for EC members, and initiated collaboration opportunities, notably the European PV conferences further detailed in Section 14.2.

The EC photovoltaic pilot program was conceived to support multikilowatt-scale projects (large at the time) during the RTD's second framework 1979–1983. From 31 tenders received in 1980, the Commission selected for support 15 projects listed in Section 13.2.

In due course, the European Commission adopted a new overarching series of 5-year Framework Programmes to support all science and innovation. The first of these ran from 1984–1988, with a budget of €3.75 billion, while the end of our time frame fell within the 5th Framework Programme 1998–2002, with a budget of €14.96 billion for all science and technology funding.

The solar program, however, suffered a severe budget cut around 1990, following a substantial decline in the oil price and resultant weakening of political support for renewables in some countries – a notable failure of the political stability, mentioned previously.

DG-XII's purpose was to support research and technology, and although innovative in interpreting its remit, it was not able to stray into market development.

The Directorate General for Energy DG-XVII was responsible for overall energy policy. As the solar energy sector matured and systems were able to viably deliver meaningful levels of energy, DG-XVII began to provide support alongside the Research Directorate.

In 1978, the Energy Directorate started its Community Programs for demonstration and promotion of energy technologies. There was a call for proposals under the program each year until 1989. In this period the Commission committed over €900 million to 1788 selected projects, of which 912 were classified as "alternative energy sources." About 100 of these incorporated PV – sometimes as hybrids with wind, solar thermal, or diesels. These projects, which were overseen on behalf of DG-XVII by Willi Kaut, had a combined capacity of nearly 1½ MW$_P$, detailed in Section 13.3, and received €20 million in support.

After the end of this program, it was replaced in 1990 by "Thermie" (European Technologies for Energy Management) established to support rational use of energy, renewable energy resources, solid fuels, and hydrocarbons. It is available for innovation, dissemination, and so-called "targeted projects."

The Directorate General for Development DG-VIII, which handled the EC's overseas aid and development, recognized the potential of solar energy for providing power in emerging economies, so it too contributed to the market for PV systems with support from the European Development Fund and elsewhere.

The Directorate supported individual projects, mainly of the type described in Sections 3.3 and 3.6, in many countries, particularly those covered by the Lomé Convention.[2] It also supported the establishment of local centers to provide information, training, research, industrial support, and engineering advice, such as the Centre Regional d'Energie Solaire (CRES), serving Cape Verde, Chad, Gambia, Ivory Coast, Mali, Mauritania, Niger, Senegal, and Upper Volta, as they were then called.

8.4.2 European Nations

The various nation-states within the European Community funded the Commission's programs, of course. They could also adopt their own national policies on renewables. There was a spectrum of differing approaches: from the progressive – those countries with a strong national program – through reactive – those who were happy to leave most support to the European Commission – to regressive – those that saw no future for PV and tried to cut the EC budget.

---

2 See glossary for the basis of this convention and the countries involved.

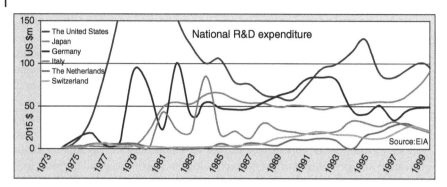

Figure 8.2 National PV expenditure by selected countries.

France, for example, qualifies as "reactive." It disbanded its national program and suggested that the European Commission take it on, complete with the man in charge, Wolfgang Palz^{qv} – which the Commission did. French companies were encouraged to benefit from European projects and obtained thereby quite a lot of PV infrastructure for France's remote regions and overseas departments. It established for a time the Commissariat à l'énergie solaire (COMES) under Henri Durand, which together with CNRS and AFME undertook various programs, including PIRDES led by Michel Rodot, but budgets were modest.[3]

Rather than flog stolidly through the programs of every other individual country,[4] let's just look at Germany, as an example of the "progressive" approach; and the United Kingdom, more in the "regressive" camp.

Germany centered its approach to the PV industry within the Bundesministerium für Forschung und Technologie (BMFT – the research and technology ministry), with Walter Sandtner in the driving seat for many years. This ministry provided practical and financial support to the German PV industry, while coordinating its activities to prevent too much duplication. It was unwilling to support more than one or two companies in any area of technology or expertise; and this doubtless influenced the number of consolidations and mergers in the German PV sector.

There was also close collaboration with the European Commission. It was rare that Germany failed to field the largest national delegation to European PV conferences, for example.

When Walter Sandtner first told his projects department about his idea for a 1000 PV Roofs Programme, "They sent me a long letter immediately afterwards with excellent arguments why such a programme could not be realised." In due course he overcame opposition from them and others to launch the program in 1990. It

---

3 For a fuller view of French PV policy, see Alain Ricaud's paper[143]; PIRDES: see glossary.
4 Though funding by Italy, Switzerland, and the Netherlands is shown in Fig 8.2.

provided subsidies of up to 70% for the installation of solar systems between 1–5 kW$_P$ on the roofs of private houses. Both the central government and the Bundesländer (federal states) participated and eventually some 2500 systems were installed.

Germany was also notably innovative in formulating policies designed to support the PV sector, as it matured. The Stromeinspeisungsgesetz (Electricity Feed-in Act) enacted in 1991 placed an obligation on electricity utilities to accept the power generated by renewable energy plants. This provided a model for "priority access to the grid" as later required by the first European Renewable Energy Directive of 2001. This act was also the first to legislate for feed-in tariffs though, as discussed in Section 13.4, the tariffs didn't really become effective at the national level until the EEG Act at the start of the twenty-first century.

The United Kingdom enjoys a very similar solar radiation regime to that in Germany but, by contrast, had very little sympathy for supporting the first solar generation.

UK industrial policy, even in the pre-Thatcher era was anti-interventionist with great emphasis on the free market. The government was ideologically averse to identifying promising technologies – which it called "picking winners" – and had little ability to do so, because its drive for "smaller government" meant that little technical expertise remained in the civil service.

It did consider on several occasions what its policy toward renewables should be (see further in Section 13.4). Not unrepresentative is its 1977 conclusion that "photovoltaic devices cannot be considered seriously as a prime source of electricity in the United Kingdom either for replacing or contributing to power in the national grid" [81].

There was sporadic involvement of British companies in European programs – its 2% share of the EC demonstration projects is typical. The country was shamed into joining the IEA activities outlined in Section 8.8, after Bernard McNelis[qv] went to inaugural meetings at his own expense. Whenever there was pressure to reduce EC renewables budgets, the United Kingdom was likely to be involved.

Latterly, the United Kingdom did "come to the party" after introducing feed-in tariffs in 2010. It must be acknowledged, with the benefit of hindsight, that there is no evidence the UK economy is harmed by this tardy entry – arguably it has ridden on the coattails of other countries spending the hard bucks.

8.5 Policy and Developments in Japan

The third powerhouse region supporting the first solar generation, Japan, adopted a different approach again. From the start, policy was focused as strongly on domestic utilization as it was on industrial development – not surprising for a country that was more than 75% dependent on imported energy sources at the time of the first oil crisis.

In fact the origins of Japan's early support program, the Sunshine Project, predate this crisis. A research proposal on a solar energy technology had been submitted by the National Electrotechnical Laboratory the previous year. At the time this was to be part of a different research project being conducted by the Ministry of International Trade and Industry (MITI^{qv}). It failed to meet the criteria for that project, which could not fund projects with longer than 10-year timescale, and PV clearly needed a longer time horizon.

MITI started to plan a new longer term program – adding three other energy technologies: coal liquefaction, geothermal, and hydrogen to bring the budget up to "critical mass." And then the first oil crisis struck. Political support rocketed and the Sunshine Project started in 1974 with a budget of ¥2.5 billion. At first, solar thermal generation was the top priority, with PV only a small part of the program.

The project started with research and development on solar cells and systems. Sharp was already involved in PV technology and its embryonic market, and other electronics companies, including Sanyo^{qv}, Fuji, Nippon Electric (NEC), and Kyocera^{qv}, joined to participate in the program. In the early 1980s, the New Energy and Industrial Technology Development Organization (NEDO) started to promote more practical research, development, and demonstration (RD&D) on manufacturing technology, cost reduction, and PV system applications.

Meanwhile the commercial market for solar-powered consumer products was starting to grow, led by calculators using amorphous silicon cells from Japanese producers. The power systems market was developing more slowly, and Hitachi, Toshiba, and NEC withdrew from PV business in the late 1980s. The annual Japanese production capacity had grown to 10 MW$_P$ by the end of 1980s.

In 1993, guidelines were agreed for the connection of PV systems to the grid, and this paved the way for the introduction, the following year, of a 50% subsidy to encourage rooftop installations. The program was open to residential homes, housing complexes, and collectives. Rooftop systems were fitted on over 500 rooftops in the first year, with a combined capacity of 1.6 MW$_P$. Deployment accelerated rapidly after that as shown by the annual figures in Section 13.5, which also gives the expenditure, average prices, and subsidy levels.

By the end of 1994, total cumulative installed capacity in Japan had topped 30 MW$_P$. Government set a 2010 cumulative deployment target of 4600 MW$_P$ (4.6 GW$_P$), and the rooftop program became the key volume driver in the Japanese PV industry, although not without casualties, as noted below.

In 1997, MITI grew the program through the introduction of the "New Energy Law" to encourage mass production, with a matching change of name from "Residential PV System Trial Program" to "Residential PV System Disemmination Program" (significant to them, even if not to you and me). Total installed capacity in Japan exceeded 200 MW$_P$ by the end of the century.

Subsequently, the rate of deployment continued to accelerate, with continuing stimulation from the rooftop program. Japan became the world's leading PV supplier early in the new millennium. It cut back support for the rooftop program in 2003, and when it ended the following year, 400,000 systems with over 1 GW_P of combined capacity had been installed.

By that time the program had claimed the scalp of Sanyo[qv] president Sadao Kondo, who resigned when the company was found to have supplied under-performing rooftop systems, and was replaced by Yukinori Kuwano[qv]. MITI also banned Sanyo from the program for 3 years, and the same sentence was handed down to Kyocera[qv] around the same time for a different misdemeanour.

8.6 Policies in Other Parts of the World

This will be a short section! Things have changed markedly since, but during our time frame there was little concerted policy toward solar electric power outside the "big three" of Japan, Europe, and the United States.

Some European countries had their own national programs in parallel with the EC's activities, described already. Although Section 8.4 considered only two specific nation-states, there were clear national activities in others, notably Italy and the Netherlands, which allocated reasonable budgets, as shown in Fig. 8.2. There were smaller programs in France, Spain, Greece, Finland, and Austria, while Denmark took a lead in introducing net metering for PV systems at a national level.

Outside the European Community, Switzerland was active in the sector, particularly for grid-connected applications, although much of this stemmed from cantonal and local initiatives, rather than the central government.

Producer nations, such as India, Brazil, and latterly Korea and China did what they could to support their manufacturing companies, although none had programs on the scale of the big three. Similarly, Australia provided support for its world-leading research center at UNSW-SPREE[qv].

Some of the countries that were not active in the industry itself none-theless introduced measures to encourage PV deployment. Canada, for example, introduced fiscal and other support measures for renewable energy deployment. In the 1990s, several African countries led by Mali, Botswana, and Ghana introduced measures to support the use of PV for rural electrification.

Subsequently, things have moved on. Most countries now have policies to address renewables in general, and in many cases PV specifically.

8.7 International Aid

This section has so far addressed the policies adopted by governments toward research, industrial development, and deployment in their own countries.

Because many of the most viable applications were off-grid, PV is particularly well suited to rural electrification in emerging economies.

Soon, therefore, governments and agencies started offering solar power in their overseas development and aid programs. In addition to the US and European aid agencies already mentioned, national governments supported PV deployment through agencies such as the French Ministre des Affaires Étrangères et du Développement International (MAEDI), Netherlands' overseas aid department (DGIS), the UK's Department for International Development (DFID), and German Gesellschaft für Technische Zusammenarbeit (GTZ, now GIZ). The French Atomic Agency Commission also supported projects in French Polynesia, led by Patrick Jourde.

International agencies also played a major part, including the World Health Organization (WHO), the World Bank[qv], and several branches of the United Nations, particularly the United Nations Development Programme (UNDP).

Providing aid in the form of technology is a very different proposition from sending food, medicine, or supplies. Some of the aid-funded PV projects did not unfortunately reflect this distinction. Few recipient countries and organizations had the resources, expertise, or funding for operation and maintenance of the systems after they were installed. Ideally, these programs should have reserved part of the budget for ongoing O&M activities. In practice few did so; donors and recipients "Just want to cut the ribbon, switch the lights on for a photo opportunity, and then [go home]." [68] All too often, the main emphasis was on the number of systems that had been supplied, with little consideration to how long they would be of service.

There were too many instances where project delays meant the batteries had died before they reached site; solar equipment was "being stolen as fast as it was installed"; and where whole systems were rendered inactive for want of one small spare part. An expert on aid applications, Bernard McNelis[qv] concludes [68] that such problems combined in many cases with excessive bureaucracy, meant that "these projects fulfilled a purpose in introducing the technology (often at huge cost), but little else."

8.8 International Collaboration and Comparisons

Having reviewed national and regional activities, we end with a brief look at international cooperation in the International Energy Agency[qv] (IEA) and elsewhere.

8.8.1 International Expenditure Comparisons

The IEA maintains data on the energy affairs of its members, enabling support given by countries mentioned above in this chapter to be compared, based on its PV RD&D spend data [108] in Fig. 8.2.

Figures for the European Commission are not available here but the data for Germany, Italy, and the Netherlands, which accounted for the lion's share of national expenditure in the European Community are shown. Outside the European Commission, Switzerland also committed significant investment later in the period as shown.

8.8.2 IEA Photovoltaic Power Systems Program

For many years, the IEA resisted giving photovoltaics its own program, although it was squeezed into the Solar Heating and Cooling Programme (SHC) from 1990 to 1995 as Task 16 – Photovoltaics in Buildings.

Finally, the IEA Photovoltaic Power Systems Programme (PVPS) was established in 1993, thanks to years of effort by Roberto Vigotti[qv], who served as its first chairman. In common with other IEA collaborative R&D agreements, the PVPS participants (subscribing member nations of the IEA) exchange information and undertake joint projects, and establish various work streams known as tasks.

Task 1 is generic ongoing work on analysis and outreach, including regular newsletters. Most other tasks carry out specific projects and then cease.

Task 2 collected and analyzed operational data from various types of PV plants around the world, to draw up guidelines on performance, reliability, and sizing of PV systems and subsystems. Task 3 was established to look at stand-alone applications in remote and island communities. Task 5[5] covered grid interconnection issues in building integrated and other decentralized systems. Task 6 ran until 1997, looking at plants for large-scale power generation. Task 7 focused on PV power systems in the built environment.

Tasks 8 and 9 are still running, working on large-scale PV power generation systems and in cooperation with developing countries, the latter originally led by IT Power[qv]. Several new tasks were added after our time frame.

8.8.3 PV Global Accreditation Programme

This important project deserves mention and is included here, although it is an industry-led rather than government-led initiative.

In the mid-1990s, Peter Varadi expressed the concern that the lack of any common approach to PV system quality and design was restricting the industry's ability to raise external finance for projects. The international standards for solar module testing were fine in themselves, but the quality of other subsystems, and of the overall design, needed to be certified too.

---

5 For some reason, there was no Task 4, as far as I can tell.

The leading industry associations EPIA[qv] and SEIA[qv] supported this view, as did the European Commission[qv], World Bank[qv], IEC[qv], and others. In 1996–1997, the PV Global Accreditation Programme (PV-GAP) was duly incorporated in Geneva, to introduce a suitable certification scheme [109], by Markus Real[qv] and Peter Varadi, its first Chairman.

Subsequently, in 2007, PV-GAP was incorporated into the IEC's system of Conformity Assessment Schemes for Electrotechnical Equipment and Components (IECEE).

9

The Next Generation

"In a civilized country when ridicule fails to kill a movement it begins to command respect."

Mahatma Gandhi [110]

Let's finish this "biography of terrestrial photovoltaics" by reviewing what the first solar generation had achieved by the end of our time frame, as the PV sector got ready to pass the batten to the next generation.

9.1   What Did the First Solar Generation Achieve?

As the twentieth century was ending, there was a slight sense of attitudes changing. The initial sheer disbelief about the prospects for solar energy had mellowed a little.

In the search for a balanced view of what the sector had so far achieved, let's put the *First Solar Generation* (*FSG*) in the spotlight, being interviewed by the renowned hard-nosed reporter *A Sceptical World* (*ASW*). To open, he softens up his subject with a compliment and an easy question:

ASW: Congratulations! I'm told you have now passed the 1 GW milestone. How much is one gigawatt? What will it power?

FSG: A gigawatt is one thousand megawatts; in other words, one billion watts of power. When we started back in 1973, the installed capacity was between 2½ and 3 MW, so we've now got 450 times as much.

What would it power? Most of our systems are for remote projects like lighthouses and radio repeaters. If they need an average of 500 W each, a gigawatt would power two million systems.

The Solar Generation: Childhood and Adolescence of Terrestrial Photovoltaics, First Edition. Philip R. Wolfe.
© 2018 by the Institute of Electrical and Electronic Engineers, Inc. Published 2018 by John Wiley & Sons, Inc.

ASW: Does that mean solar is no good for ordinary people? I haven't got a lighthouse or radio repeater. The only solar products I've seen are gimmicks.
Couldn't I use solar panels on my house?

FSG: Oh yes, solar panels are now quite widely used for residential power. There are about 70,000 solar rooftops installed around the world mainly in Japan, the United States, and Germany.

If you want to know what 1 GW represents in those terms: It's the equivalent of an average nuclear power station. One gigawatt installed in Europe would deliver enough power for about a quarter of a million households.

ASW: Is that all! That's barely one tenth of 1% of the world's electricity, and it's taken 26 years. If you carry on for another century, you'll only reach ½%.
Do you never expect solar to meet a serious share of global electricity?

FSG: It's not right to project trends that way. Our growth is not linear; it is not a straight line from 3 MW to 1 GW over 26 years. No, our expansion has been exponential, with yearly installations rising from under 1 MW in 1974 to nearly 200 MW in 1999. That's a compound annual rate of 25% per annum.

If that continues, we would reach 44 GW in 2015, over 400 GW by 2025, and 120,000 GW by 2050. Those figures would represent over 4% of current global electricity in 2025, and enough to meet the world's power needs ten times over by 2050.

Sure, in practice, capacity growth will slow down before then, but the short answer to your question is, Yes, we expect solar to make a huge contribution to world energy during this century – probably the biggest source of power, in fact.

ASW: That's one hell of a prediction!
How good have your forecasts of future growth been up to now?

FSG: We have to admit they've been variable. We usually hit the forecast – but often rather later than we had predicted!

ASW: I know most of your installations are subsidized. If your forecast is actually right this time, the subsidy cost would become astronomic.
Why should the world carry on subsidizing solar energy year after year?

FSG: The quick answer is, it shouldn't – incentives are needed only to kickstart the market. All energy sources have been subsidized. As you know, expenditure on nuclear power has been enormous, and the fossil fuel industry gets subsidies eight times higher than the entire renewable energy sector [111].

The latest incentive mechanisms, such as feed in tariffs, to be introduced in Germany soon, allow the level of subsidy to decline as the market takes off.

ASW: Yes, but surely solar power needs high subsidies in order to compete. Won't the industry die again as soon as the incentives are removed?
Isn't solar just too expensive to compete the free market?

FSG: That is all changing too, thanks to an even more important milestone we've achieved while expanding our capacity.
We have brought solar module prices down under $5; that's less than 1/40th of what they were when we started 25 years ago. If we do the same again, photovoltaics will be the cheapest source of energy on the planet.

ASW: Sounds good, but surely you've now had most available cost savings.
Solar panel prices can't carry on coming down for ever can they?

FSG: We used to believe they couldn't; we thought we could get costs down to $2/W, maybe $1/W, and then we'd have to find new cheaper solar cell materials. Now we're not so sure. Recent studies have shown that existing technologies can get down way below $1/W.
In fact, if you look at what's happening in the computer industry, their costs keep going down and down, as the volume increases; there doesn't seem to be any bottom limit. Fundamentally, solar cells are semiconductor devices; we're starting to believe the same can happen here.

ASW: Your market is exploding, costs are tumbling, and you're subsidized.
Are all you solar power guys rich then?

FSG: Sadly not. The problem with manufacturing industries like ours is that the faster you grow, the more working capital you need. A lot of this investment has come from the entrepreneurs in the sector themselves – people have been getting poorer rather than richer. But it's all in a good cause!
Our industry will continue to swallow huge amounts of cash while it is expanding. Don't expect to see solar power billionaires any time soon.

ASW: You said most of the rooftop systems are in Japan, Germany, and the United States. Is that related to the support and subsidy systems they have?
Which national policies have worked best for the solar industry?

FSG: With the benefit of hindsight, policies which generated volume for the manufacturers have been most successful, at least from an industrial perspective. The US Block-buys and the Japanese rooftop program created real volume for manufacturers and so brought costs down.

Support for research and for pilot and demonstration programs was good for advancing knowledge and for academia, but on its own didn't generate world-leading companies.

Even the German program has been quite modest so far, however, they are talking about great things over the next decade. That's why the leading companies have always been Japanese or American; the only time European companies got to the top, was when they acquired US businesses.

ASW: You sound happy with cost and markets trends. What about technology?

Do you predict improvements to cell efficiency and performance too?

FSG: That's an area, unlike cost, where there definitely are practical limits. Single-junction silicon cells dominate the market, and their maximum achievable efficiency has been calculated at 33%.

Our best researchers have pushed laboratory efficiencies up to 25% – having started at about 14%. They will keep moving forward, but it will get ever harder as they approach the theoretical limit. Production solar cells typically lag laboratory efficiencies by about one quarter, so there's still scope for manufacturers to work on closing that gap.

More sophisticated types of cell have higher theoretical efficiencies – over 80% in the most extreme cases. The industry is working on many new cell types, so we expect to see efficiency continuing to inch upwards.

Having said that, there has been far more focus on reducing cost than on improving performance, and we expect that priority to continue.

ASW: Yes, I've heard about some of these new thin film and organic materials. People are saying they will be the next big thing.

How fast do you expect new technologies to enter the market?

FSG: That's a very difficult question – and you probably shouldn't believe our answer anyway! We have got that wrong in the past.

We used to believe that new lower cost technologies, such as amorphous silicon for example, would take over the market, progressively replacing mono- and polycrystalline cells. In practice, we were wrong. The original technology has been very successful in reducing cost and increasing efficiency, and so has retained the lion's share of the market.

Several of the new materials look promising, but the jury is out on when and how much they will penetrate the market.

ASW: Well that's it; thanks for being so candid. I suppose we have to wish you good luck for the future. If you're actually right about the direction of travel, I guess we'll be seeing a lot more of you. Perhaps we'll even start to see solar panels on rooftops all round the country – that'd be quite something!

9.2 A Time Traveler's View

Throughout the book, I have tried to be true to the time frame selected at the start, 1973–1999. Accordingly, the responses in the interview above are based on the situation and information available at the end of the last century.

It is hard, however, to completely ignore that I am writing this more than 15 years later. If the interviewee had the ability back then to travel to the present time, he or she could have refuted the skepticism more forcefully.

Take market growth, for example. Many were doubtful that the historical 25% compound growth rate would continue to reach 44 GW by 2015. In fact, growth accelerated to nearly 40% compound in the first 15 years of the new millennium. Actual capacity at the end of 2015 was over 200 GW.

Have our forecasts improved? Actually they have, in fact, many projections have been proved conservative. In estimates published by PV Systems^{qv} in 1984, for example, it forecast [112] market volumes for 1990 and 1995, which proved overoptimistic by 9 and 7 years, respectively. Maycock's 1994 projections [113] by contrast were too cautious. He was spot on for the year 2000, but his 2010 figure was met 6 years early in 2004.

Observers outside the industry have joined in. Speaking in 2016, the futurist Ray Kurzweil said [114]: "In 2012 solar panels were producing 0.5% of the world's energy supply. Some people dismissed it, saying that's nice but at a half percent solar is a fringe player, not going to solve the problem. They were ignoring the exponential growth – like they ignored the exponential growth of the internet and human genome project. Half a percent is only 8 doublings away from 100%. Now 4 years later solar has doubled twice again. Now solar panels produce 2% of the world's energy, right on schedule. People dismiss it saying 2% is nice, but a fringe player. That ignores the exponential growth – which means it's only 6 doublings or 12 years from 100%." I have yet to hear an industry insider suggest we'd supply the global electricity market that soon!

What about cost reductions? They have indeed continued along the 81% progress ratio line. The $1/W_P$ level (in 1980 dollars), we used to think was the bottom limit for crystalline silicon, was breached back in 2009. We are now looking at fractions of $1 even in today's dollars.

As far as politics is concerned, our observations on the best support frameworks have been borne out in practice. The introduction of feed-in tariffs stimulated explosive expansion, and the resulting cost reductions. In fact, they were so successful, regulators had to introduce the degression mechanism to ensure that tariffs declined as fast as prices. Most of the grant-based incentives have gone now, and new measures – such as tax credits, obligations on utilities, and tender purchases – are all designed to stimulate deployment while containing expenditure.

Solar cell efficiencies have continued to rise. Panasonic's heterojunction cell overtook monocrystalline silicon at an efficiency of 25.6% in 2014, an improvement of almost one-fifth in a decade. Amongst thin films, both cadmium–telluride and CIS have improved strongly since 2010.

Organic technologies, such as dye-sensitized cells, continue to progress. New derivatives, which hadn't even been thought of during our time frame, such as quantum dots and perovskites, make even more dramatic advances. I still think the jury is out on whether these materials will reach the mass market.

But don't take my word for it – I thought amorphous silicon would be ruling the world by now!

9.3   So Where Do We Go Now?

The terrestrial photovoltaics sector faces new and different challenges as it leaves the adolescent years for young adulthood.

The disbelief and cynicism we faced has receded. Many people remain skeptical, but we now have a growing army of active supporters, like Ray Kurzweil (quoted above).

The early driver of energy security remains extremely prominent – and has been joined by a further strong stimulus, climate change. The major obstacle, high cost, is rapidly being overcome.

As I write, the Second Solar Generation is already in the driving seat. The global PV industry is immensely larger than that which ended the last century. We go to solar expositions these days and find more PV capacity in the exhibition hall than the entire world's annual production at the start of our time frame.

The second generation is also much more professional. By comparison, we were just amateurs – perhaps not surprising as we were trying things no one had done before. The first solar generation looks on their accomplishments with pride, but nostalgia too – are they having as much fun?

Where will it end? Some of the early pioneers thought that PV would always be restricted to off-grid and specialist applications [86]. But the majority saw beyond that to a serious contribution to global energy. For the first solar generation, this expectation was mainly blind faith. For our successors, it is now a confident prediction. Unless someone invents a cheap small-scale nuclear fusion generator pretty damn soon, solar photovoltaics will become a leading – *the* leading – energy source for the planet.

I look forward to reading about it when someone writes the sequel.

Part II

Encyclopedia – People, Organizations, Events

> *"I'd put my money on the sun and solar energy.*
> *What a source of power!*
> *I hope we don't have to wait until oil and coal run out . . ."*
> Thomas Edison – in conversation with Henry Ford
> and Harvey Firestone, 1931 [115]

One of the many features that attracts me to Chaos Theory and the Mandelbrot Set (see Fig. B.1 on page 314) is that you can focus ever more closely on individual details, and still see versions of the whole picture.

You have now read Part I and gained a picture of the formative early years of terrestrial photovoltaics. We have explored the broad themes of industry, technology, markets, politics, and so on.

Now in Part II, we will look at the same picture from a different perspective, reviewing the contributions made to the sector by individual people, companies, research centers, administrations, and events.

The Solar Generation: Childhood and Adolescence of Terrestrial Photovoltaics, First Edition. Philip R. Wolfe.
© 2018 by the Institute of Electrical and Electronic Engineers, Inc. Published 2018 by John Wiley & Sons, Inc.

10

Who's Who: Profiles of Early PV Pioneers

The reasonable man adapts himself to the world:
 The unreasonable one persists in trying to adapt the world to himself.
Therefore all progress depends on the unreasonable man.
George Bernard Shaw [116]

Too many books, in this sector and others, give the impression that inventions, companies, or events are the brainchild of just one individual. In practice, this is rarely, if ever, the case. The most successful organizations are built on groups of good people. All the pioneers I interviewed acknowledged the contributions of the teams around them. The following Who's Who is therefore an inevitably incomplete record of key people active in the early PV sector.

10.1 Inclusions and Omissions

I have tried to be objective and consistent, and have consulted others, but it's impossible to escape all subjectivity. My background makes it likely that industrialists may be overrepresented compared to researchers, politicians, consultants, or public servants.

Any omissions and errors are entirely mine. The criteria I have sought to adopt are as follows:

For most of the people listed, terrestrial solar photovoltaics was the primary focus of their work. Individuals for whom PV was only a small part of their career are included only in exceptional circumstances. The list shows people who were active for a significant proportion of our time frame. People whose main contribution to the PV sector came after the end of the twentieth century are therefore not included.

Organizations are made up of people, all making some contribution to the overall outcome. The following list has adopted the "buck stops here" approach, and singles out the founders, leaders, and principal board members of relevant PV organizations or departments.

The Solar Generation: Childhood and Adolescence of Terrestrial Photovoltaics, First Edition. Philip R. Wolfe.
© 2018 by the Institute of Electrical and Electronic Engineers, Inc. Published 2018 by John Wiley & Sons, Inc.

All in all, this selection policy means that many significant contributors are not individually listed below, because they were in middle management, made their main contribution later, or were simply less visible. Some are recognized in Section 10.3, and you should refer to the full index in Section G of Part III for a more comprehensive list of those who contributed to the early development of terrestrial photovoltaics.

The almost complete omission of women is, believe me, not a matter of editorial selection policy, but merely a sad reflection of the gender gap prevalent at senior levels of the technology sector at the time.

10.2 Profiles of Selected Terrestrial PV Pioneers

This list contains just a selection of the key PV pioneers (see also Section 10.3) involved in terrestrial PV in the first solar generation.

Remember: Our period covers primarily the period 1973–1999; later activity is mentioned only in summary as "latterly" or "subsequently" (and italicized in timelines). Superscripted "qv" relates to other people, organizations, events, or awards profiled in this book.

Where possible, photographs date back to, or close to, this time frame. Where I have used academic titles, they are those applicable in that era. Some company titles are abbreviated to acronyms, but you can look up the full names – if you really want – in the Glossary section.

Dr. Tapio Alvesalo

One of the first Scandinavian scientists to recognize PV's potential, Tapio Alvesalo led the region's leading early solar business (Fig. 10.1).

Alvesalo obtained his physics masters at Helsinki University of Technology, where he stayed for his doctorate, working on quantum liquids at ultralow temperatures. He moved to the United States for 3 years from 1975 as a research associate at Cornell University's Laboratory of Atomic and Solid State Physics, with the subsequent Nobel Laureate Prof. Robert Richardson.

On returning to Finland in 1979, Dr. Alvesalo joined the Finnish national oil company Neste. He was appointed manager of research and development and a member of the "Phi" group, which new chief executive Jaakko Ihamuotila had established in the

Figure 10.1 Tapio Alvesalo at EPIA event [68] showing his talents extend beyond physics.

wake of the oil crises to evaluate emerging energy technologies. Its work led to Tapio's appointment as head of strategic R&D at Neste's Battery Division in 1983.

Tapio recognized the synergies between the nascent PV technology and this division's activities in advanced energy storage, hydrogen, and electric vehicles. He proposed a new corporate business venture, Neste Advanced Power Systems (NAPS[qv]), which was launched in 1986. Alvesalo remained chief executive of this business until 1998, and then served as chairman until he retired.

The PV business grew strongly under his leadership, establishing overseas ventures and selling globally. Alvesalo presided over the creation within Neste of Microchemistry and the acquisition of Chronar France[qv].

He was also active in developing the technology and PV sectors beyond his own company. Alvesalo served on the board of EPIA[qv] for many years and as its chairman twice; he chaired the Finnish National Research Program on Advanced Energy Systems and Technologies for 5 years. He was knighted by the President of the Republic of Finland in 1993.

Subsequently, after Neste and the Finnish national utility company IVO merged to form Fortum in 2000, Dr. Alvesalo was appointed Vice President of Corporate Technology, and presided over the sale of NAPS. After his retirement in 2004, he directed Finland's Millennium Prize Foundation and latterly serves as board member and partner in a new Finnish PV company, Solnet.

Peter Aschenbrenner

Peter Aschenbrenner has enjoyed a varied career in the PV industry, spanning the range from engineering to sales with equal intellect, thanks maybe to his unusual degree majoring in both engineering and studio art.

Aschenbrenner joined Arco Solar[qv] directly from Stanford University[qv], where one of his fellow graduates was Kari Yerkes, Bill's daughter (Yerkes[qv] called her his "chief talent scout"). His first role was as an applications engineer developing water pumping and DC home systems for developing countries.

In 1980, he was sent to lead technical support activities at Arco's new European office – at that time in Milan – working with Franco Cugusi and Achille Manfredi, both seconded from local partner Microelit. In 1983, Aschenbrenner and Maurizio La Noce moved the European office to the United Kingdom, near London.

Aschenbrenner returned to the California headquarters the following year working in product development, this time on amorphous silicon cells. When Arco signed their cooperation agreement with Siemens^{qv}, he moved to Germany as comanager of the joint venture PV Electric alongside Hubert Aulich^{qv}. A key focus was semitransparent thin film solar sunroofs for cars.

Aschenbrenner again returned to the United States in 1990, and 3 years later left Arco Solar to join AstroPower^{qv}, as head of sales and marketing. He was one of the senior team who handled the successful flotation of Astro-Power in 1998.

Subsequently, Peter Aschenbrenner moved back to California to join Sun-Power^{qv}, where he is vice president of corporate strategy and business development.

Dr. Hubert Aulich

Hubert Aulich is one of the longest serving solar specialists in Germany and an expert on PV materials.

He obtained his doctorate in Physical Chemistry in the United States from New York University. In 1974, Dr. Aulich joined the Central Research Laboratories of Siemens^{qv} in Munich, working initially on fiber-optic communication and then on photovoltaics.

From 1988, he was managing director of PV Electric, Siemens joint venture with Arco Solar^{qv}, where he worked on commercialization of amorphous silicon thin film technology with Peter Aschenbrenner^{qv} and others.

In 1992, his responsibilities extended to crystalline silicon products too, as managing director at Siemens Solar in Germany. Later he was appointed senior vice president for technology and research for Siemens Solar Group, managing all thin film and crystalline silicon research and development and systems activities in Germany, the United States, and Japan.

Hubert left Siemens in 1997 to cofound PV Silicon with Friedrich-Wilhelm Schulze. When this later merged with Crystalox^{qv}, Hubert managed German operations and was subsequently appointed to the board. He also chaired the collaborative industry-research project SolarValley Mitteldeutschland.

Latterly, Aulich has served on the board of trustees of Fraunhofer ISE^{qv}. He participated in the subsequent MBO in 2013 of the former Crystalox^{qv} plant in Bitterfeld, which produces polycrystalline silicon using the Siemens Process. He then established Sustainable Concepts to provide solar-powered water purification for developing countries.

Dr. Allen Barnett

Allen Barnett is one of a handful of academics to have also founded significant industrial businesses in the sector.

Barnett earned his bachelors and masters degrees from the University of Illinois and doctorate from Carnegie Institute of Technology, all in Electrical Engineering, before going into industry at General Electric. He then founded his first commercial company, Xciton, which used semiconductor materials technology in the LED and information display market. The company was sold to National Semiconductor 5 years later.

Dr. Barnett then joined academia at the University of Delaware, working on the application of advanced materials science and technology to the development of new electronic devices. He was director of the Institute of Energy Conversion[qv] from 1976–1979, and then professor of Electrical Engineering until 1993.

From 1983–1989, Barnett also served as general manager of the AstroPower Division of Astrosystems, and he became chief executive when AstroPower[qv] became an independent company. He worked exclusively at AstroPower from 1993, and led it through a successful flotation in 1998.

Latterly, in 2003, he returned to the University of Delaware and led the US "Very High Efficiency Solar Cell Project," before joining Martin Green's team at the University of New South Wales[qv], as part of the US–Australian program.

Allen has served on many advisory boards, including SEIA[qv] and the US Department of Energy's National Center for Photovoltaics, based at NREL[qv]. His honors and awards are also too many to list, including the William Cherry Award[qv] and the Karl Böer Medal[qv]. He personally has been a prominent supporter of Mitochondrial Research.

Joachim Benemann

Joachim Benemann is one of several of the early PV pioneers who came from the nuclear industry, having joined the Siemens subsidiary Interatom in 1965.

Returning in 1982 from a nuclear research posting in Indonesia, Benemann was asked to take over the solar activities of Interatom's new technologies division. He agreed even though he "had never even heard of photovoltaics" [69].

Two years later, the glass company Flachglas needed someone with solar expertise to exploit the nascent concentrated solar power market and recruited Benemann to lead its solar business. At the time this CSP sector – where the sun's heat rather than light is used to generate power – was much bigger than PV, especially in California where multi-megawatt plants were being developed.

Benemann soon added a PV business stream to Flachglas Solartechnik^{qv}, focusing initially on solar façade applications. At this time he also took the company into EPIA^{qv}, becoming one of the founder directors. He progressively broadened the company's PV activities, as more fully described in his chapter of the *Solar Power* book [43], diversifying into integrated manufacturing, in partnership with Shell^{qv}, and control electronics through the partial acquisition of SMA^{qv}.

Subsequently, when in 2000 the parent company Flachglas AG decided to concentrate on its core business, Joachim Benemann was involved in the management buyout of the solar business. He retired a few years later, about 18 months before the business was broken up.

Dr. Elliot Berman

No one outside Japan has been working in the terrestrial PV business longer than Elliot Berman; he started his research in 1968 – before even this book's starting date for "the early PV era."

Having won his doctorate as an organic chemist at Boston University, Dr. Berman's first job was with NCR (National Cash Register) in Ohio. He worked on carbonless paper, early computers, and photochromic materials. The optics company Itek (so called allegedly by a founder who had moved from Kodak saying "I took Eastman Kodak") was interested in his photochromic work and gave him a job back in Boston.

He tired of research into nonphotographic copying and, when asked in 1968 what he'd rather work on, he picked out solar power. This was well before solar had any attention, but Berman felt it had potential for the developing world and might make up for the fact that he "couldn't join the Peace Corps because I had a wife and kids" [117]. His enlightened boss agreed to let him work on it for a year, after which Itek would take it on if they were interested; and if not, he'd leave.

A year later Berman left, self-funding his solar cell research in rented lab space at Boston University; he established Solar Power Corporation. He started looking for a financial backer, believing his own money would run out in 9 months. In the end, it had to stretch for 2 years, until one of Itek's funders introduced him to Exxon Enterprises.

Exxon agreed to provide $2.5 million needed for his proposed 2-year research project. Although never an Exxon employee himself, he hired in the early 1970s a dozen researchers on Exxon's payroll based at their research laboratory in New Jersey. They started working on organic solar cells, improving efficiency 10-fold – but still to just 1%, because the materials were poor conductors. After 6 months when Exxon asked what the business model would be even if the research succeeded, the team decided to move to more mainstream devices and studied the silicon solar cell market. Berman recalls visiting the companies active at the time: Spectrolab^{qv}, Applied Solar Energy Corporation, Sharp^{qv}, and Philips RTC. Their focus was predominantly on space cells where cost was less important.

Berman therefore initiated a program to develop lower cost silicon cells, and this in turn launched Solar Power Corporation^{qv} into the market. After an exciting start, Exxon became increasingly focused on profitability. When in 1975 it refused Berman's request to allow a loss of $250,000 to fund further research, he left and went back to Boston University, where he started a Center for Energy Studies. He also initiated work on luminescent concentrators.

Finding life in academia rather restrictive, he jumped at the opportunity to rejoin industry when Arco's Dick Blieden showed interest in his luminescence work. Berman therefore moved to California, as Chief Scientist at Arco Solar, but fulfilling other roles including at one time marketing for rural programs – making up in the end for having missed out on the Peace Corps.

After leaving Arco, he moved back to the East Coast. He subsequently held senior positions at Photox and Zentox where he is still active.

Prof. Werner Bloss

Werner Bloss was a greatly respected academic and a charming man. Having studied Physics at the Universities of Tubingen and Stuttgart, he became a research assistant at the latter, delivering his doctoral thesis on a thermionic converter that he had developed. He took an early interest in energy conversion and wrote a textbook on Electronic Power Converters [118], which remains a fundamental reference.

In the late 1960s, Bloss was appointed Associate Professor at the University of Gainesville in Florida for 2 years. Soon after returning, he became director of a research group at the University of Stuttgart^{qv}, which he renamed the "Institute of Physical Electronics" (IPE).

Bloss was one of the first to realize the implications of the first oil crisis and was in the forefront of a successful campaign to convince ministries and other decision makers in Germany to support photovoltaics for terrestrial applications. This, and subsequent funding from the European Commission, provided the resources to establish IPE as a leader in research on photovoltaic thin films, which he considered to be the only chance for large-scale use of photovoltaics. IPE had major programs in copper sulfide/cadmium sulfide, amorphous silicon, and later copper–indium selenide (CIS).

Bloss maintained good connections with industry, and was happy to transfer technology. He was a prominent supporter of solar hydrogen technology active in Hysolar – a collaborative project between Saudi Arabia and Germany – and was involved in founding the Center for Solar Energy and Hydrogen Research Baden-Würtemberg (ZSW) in the late 1980s serving as Chairman.

Bloss was not simply a theoretical physicist; he was also active in applied technology including power conditioning, PV systems modeling, concentration, tracking, and storage. He also undertook advisory work and collaborative projects in Germany, the European Community, and further afield. Several researchers from India in particular studied under Prof. Bloss.

He was a consultant editor to *Progress in Photovoltaics* in the years before his untimely death in 1995. Werner Bloss was given the Solar Award of the German Section of ISES in 1989 and the Becquerel Prize[qv] in 1991; he was honored in 1990 by the First Class Order of the Federal Republic of Germany.

Prof. Karl Wolfgang Böer

One of the most prominent early researchers in thin film photovoltaics, Karl Böer also had a keen appreciation of the importance of real-world applications.

Böer's interest in science dates from a very early age when given a chemistry set by his father. He obtained his doctorate in Berlin in the 1950s and, when the Berlin Wall was built, emigrated to the United States, where he adopted the name Karl "because Americans can't pronounce Wolfgang properly."

After a spell at New York University, he was appointed professor at the University of Delaware, initially in physics, then physics and engineering, and eventually physics and solar energy.

Böer had been one of the first to identify the potential contribution of solar power, even before the first oil crisis [19]. He led research on solar energy technologies, focusing on copper sulfide/cadmium sulfide (Cu_2S/CdS) thin film cells, and was the founding director of Delaware's Institute of Energy

Conversion^{qv} in 1972. The next year he initiated the construction of an experimental house called Solar One, an example of Karl's recognition that good research is of little use on its own; it needs to be introduced to the wider world.

At about the same time, he founded SES^{qv} with investment from Shell to develop cadmium sulfide solar cells commercially. In 1975, he returned to academic duties at the university, after Shell took full control of SES; he said, "I have some warped genes; I can't work for other people" [119]. This attitude also led Böer to establish the American Solar Energy Society (ASES), breaking away from the international society ISES.

Karl has hundreds of academic papers, many patents, and several books [120]. The University of Delaware awards the Karl Böer Solar Energy Medal^{qv} to recognize other contributions and honor Böer for his early impact in the sector. He himself has won many awards, including the William Cherry Prize^{qv}.

Karl continues to write from his home in Florida, and participates with his wife Renate in the Böer Medal. A more detailed account of Karl's life and contributions to science is given in his book [121] and his Cherry Hill interview [103].

John Bonda

John Bonda was, in his time at EPIA, the most visible and much appreciated representative of the European PV industry (Fig. 10.2).

John's varied career started in international trade, based in Belgium. He had represented Intersolar Group^{qv} in West Africa, so when the European Photovoltaic Industry Association^{qv} was seeking its first executive appointment in Brussels, Philip Wolfe^{qv} – then an EPIA director – approached him. John took the role, initially on a part-time basis. The Association flourished under his stewardship and he was appointed Director General.

John was a master of many languages, and well-connected within the European Commission, where he soon established a close rapport with Wolfgang Palz^{qv} and others. But many EPIA members will remember him particularly for his talent for selecting meeting venues. The association became famous for the succession of castles, hotels, restaurants, and palaces where its AGMs and meetings were held.

After his death in 1999, EPIA inaugurated the Bonda Prize^{qv} in memory of one of the PV sector's best loved characters.

Figure 10.2 John Bonda on site in Egypt [68].

Dr. Dieter Bonnet

Dieter Bonnet is credited with fabricating the first viable cadmium telluride (CdTe) solar cell, and went on to found the company Antec Solar.

He gained his degree and PhD at Frankfurt University and started his career at the Battelle Institute. In 1970, he developed the world's first CdTe/CdS thin film solar cells and in 1972 his team achieved an efficiency of 6%. In 1993, when Battelle was closed, he bought out the cadmium telluride technology and created the firm Antec Solar in Thuringia to develop it. Following its subsequent closure, some of the management established CTF Solar in 2007 to continue work on the technology. Dr. Bonnet's work also contributed to the successful mass production of CdTe by First Solar.

Dr. Bonnet latterly received the Becquerel Prize[qv] for his pioneering work on II–VI compounds. He initiated Solarpact, a global network of research and industrial groups working on CdTe in 2016, the year before his death.

What Bonnet called "a semiautobiography" can be found in his chapter in the *Solar Power* book [43].

Jim Caldwell

James H. Caldwell, Jr., joined Atlantic Richfield initially as a chemical and process engineer, and had transferred to corporate planning by the time of the second oil crisis. The company's most significant potential diversification at the time was an oil shale development. Following Caldwell's study to assess this multibillion dollar investment in what would be "the world's biggest underground mine coupled to the world's largest oil refinery" [75], Arco decided it needed a partner. It approached Exxon, who eventually decided to buy the entire project, providing Arco with a substantial cash pile and leaving Caldwell free to follow his boss Ron Arnault into Arco Ventures

He was assigned to Arco Solar[qv], working at first on their utility-scale projects at Lugo and Carrisa Plain. He subsequently took responsibility for manufacturing and engineering activities, and was later appointed chief executive in 1984, with Ron Arnault moving back to a senior role at Arco corporate.

This was an exciting period for Arco Solar with a dominant position in the crystalline silicon market and ambitious thin film cell development plans. When Atlantic Richfield decided to sell the solar business at the end of the 1980s, Caldwell was keen on leading a management buyout and left the company to put together his proposal. He obtained backing from a glass company, but Arco eventually decided to sell to its joint venture partners Siemens.

Caldwell went on to establish a development business for international power and desalination projects. He subsequently went into the wind power sector with the American Wind Energy Association, worked on renewables for the Los Angeles Department of Water and Power, and undertook a project to evaluate a low carbon grid, working with NREL[qv] and the US Energy Information Administration.

He also produces wine from a small vineyard at his home in Northern California.

Dr. David Carlson

Dave Carlson is delightfully unassuming for someone with his unrivalled scientific track record in the PV industry.

After obtaining his doctorate, Carlson joined the US Army Signal Corps and soon found himself in Vietnam as a captain in charge of a communications site. Returning in 1970, he joined RCA laboratories and started work on plasma deposition of thin film solar cells, although it was not at the time a focus for them.

Dr. Carlson sounds surprised [63] when recalling that "the very first cell actually worked." He had expected that the process would produce a polysilicon thin film, but when coworker Chris Wronski characterized the device, he found it had unexpectedly good response to blue and green light, but poor output under red. The materials were sent away for analysis, and the answer came back: "It's not crystalline – it's amorphous."

The efficiencies at the time were very modest – initially less than 1%. When asked why RCA should continue the research, especially given that Paul Rappaport[qv] apparently believed it had little promise, Carlson countered "Paul's wrong." They stuck with him and were rewarded with increasing efficiencies, eventually over 5%.

During the 6 years in the late 1970s and early 1980s when Carlson headed up the PV device research group, RCA[qv] were the first industrial team to produce, characterize, and patent amorphous silicon solar cells. He is named as inventor on the patent [62], and the Staebler–Wronski effect was named after two researchers on his team.

When this work stream was acquired by Amoco's subsidiary Solarex[qv], Carlson established and led its thin film division in Newtown Pennsylvania to commercialize amorphous silicon solar cells and thin film transistors for liquid crystal displays. The division later broadened its research into copper–indium–gallium diselenide (CIGS) solar cells.

Dr. Carlson was in due course appointed Vice President and Chief Technologist of Solarex, becoming Chief Scientist of BP Solar for 13 years after BP acquired Amoco. Subsequently, BP Solar was closed, and Carlson's desire to transfer their research programs to others was not sanctioned. He continues to advise organizations in the sector. Dave Carlson's internal, national, and international honors are too numerous to list, including the William Cherry Award[qv] and the Karl Böer Medal[qv].

Günther Cramer

Günther Cramer, although not expressly a PV pioneer, merits inclusion because of his role in establishing the inverter company SMA and his adherence to his personal motto: "Let's be realists – and attempt the impossible."

Cramer had been interested in renewable distributed energy from his student days at the University of Kassel[qv], and in 1981 he cofounded the university spin-out company, which later became SMA Solar Technology[qv].

He served for many years as its CEO and later as Chairman of the Supervisory Board: the company was recognized not just for its products but also gained recognition as an outstanding employer. An experienced sailor himself, Cramer instituted an engineers' annual sailing trip for his employees.

He served latterly as the President of the German Solar Industry Association BSW-Solar. In 2012, the German President Joachim Gauck presented him with the German Federal Environmental Foundation Award.

Günther Cramer died in 2015 at the age of 62.

John Day

When John Day III established Strategies Unlimited[qv] in 1979, he perhaps didn't realize that he was creating what would become the almanac for the solar manufacturing industry. Working at the time in marketing for a semiconductor company that "competed with Intel and lost" [42], John recognized the potential of the photovoltaics sector and set up his market research business in Mountain View near San Jose in California.

John was a fixture at most of the early conferences (of course, all his clients were there), but not a mercenary presence; he was always happy to discuss developments in the sector with customers and nonclients alike. It is John who recorded [77] the near stranding of visitors to the Kythnos plant at the 1983 Athens conference[qv].

Having brought Bob Johnson[qv] and later Paula Mints into the Strategies Unlimited team, John took a less active role in photovoltaics and concentrated on Strategies' optoelectronics business. He sold the company in 2001 and retired.

Robert de Franclieu

Robert de Franclieu was working on high-voltage equipment at Alsthom when in 1981 his colleague Claude Remy[qv] invited him to join Photowatt[qv], a recent start-up within Alsthom's parent company CGE where Claude had just been appointed general manager. When Philips RTC invested in the company, production was moved to Philips premises in Caen, while Robert headed up international marketing based in the Paris suburb of Rueil Malmaison.

Initially, it was an exciting time of high growth with good support from the European Commission, national agencies, and Photowatt's substantial parent companies. However, when CGE was privatized and decided to divest to Chronar France[qv] in July 1987, many of the management lost confidence that the change of ownership was in the best interests of the company. In October, Robert declined the chairmanship and left a few days after Remy, joining him at CGE subsidiary SAFT, even though (unlike Claude) he had no ongoing contract with them.

The two worked together, with support from the government and others on a plan to buyback Photowatt. This was fraught with problems; both sides had powerful industrial backers, and it became a dogfight between the right and left of French politics. Robert describes [122] how they were still short of a quarter of the necessary funding the day before the deadline, but miraculously received a last minute check for the difference with no strings attached. Despite the challenges, their newly established vehicle called "Solar France" succeeded in buying Photowatt back from Chronar France the following year.

De Franclieu moved south after Shell's 1991 investment enabled Photowatt to establish new facilities near Lyon. He remained with the company for 3 years after it was sold to ATS. During this time, following an introduction by ATS's Klaus Woerner, he mentored a young Chinese-Canadian, Shawn Qu, who went on to found a leading twenty-first century PV company, Canadian Solar.

Subsequently, after leaving Photowatt, Robert established Apollon Solar, a silicon processing research company, from which he has since retired.

Dr. Emmanuel Fabre

Emmanuel Fabre is a soft-spoken Frenchman, who spent his entire career in the solar industry, covering many roles from research to marketing and working for several companies without ever leaving any of them!

After graduating from Ecole Centrale de Paris, Dr. Fabre joined the research department of Philips subsidiary RTC (Radiotechnique-Compelec) in 1969 and later spent a year at Philips' laboratories in New York State. While working on semiconductors on his return to Paris, he soon established contact with the team producing solar cells in small quantities at the production plant in Caen. From 1973–1974, most of his time was spent on photovoltaics research.

When Philips RTC's solar business was merged into Photowatt in 1981, he stayed with it, becoming one of Photowatt's first employees as Scientific Director. In this role Emmanuel Fabre was involved in many collaborative projects in Europe, and became the company's representative at EPIA[qv], where he was a founding director and twice served as president.

Fabre stayed with Photowatt when it was acquired by Chronar[qv] in 1987, and was appointed chairman soon afterward. He remained with Chronar France as chief executive, when Photowatt was bought back by its previous management the following year. The collapse of Chronar Corporation left the French company exposed, and a buyer was needed. His codirector at EPIA, Tapio Alvesalo[qv], was interested and NAPS[qv] duly acquired the company. Dr. Fabre became managing director of NAPS France, a role he retained until he retired in 2007.

Subsequently, Emmanuel continued in a consultancy role for a couple of years, and settled in the South of France.

Hans-Josef Fell

German politician, Hans-Josef Fell doesn't meet our criteria for this section, but is included because of his pivotal role in introducing the feed-in tariffs.

He was interested in renewables, even before joining the Green Party in the early 1990s. Fell believed the best approach was to create a commercial market, rather than the previous research and grant-based model; he thus drafted the Renewable Energy Sources Act (EEG) to guarantee feed-in tariffs for electricity from biomass, wind, and solar.

Thanks to support also from Hermann Scheer[qv], the EEG was enacted in 2000 in the face of strong opposition.

Fell is profiled in Bob Johnstone's book [82], and his enlightened approach to converting the skeptical political classes is outlined in his chapter of the *Solar Power* book [43].

Dr. Charlie Gay

Summarized timeline

1974 Joins Spectrolab

1977 Joins Arco to head R&D

1989 Appointed Arco president, stays when:

1990 Siemens acquires Arco Solar

1993 Leaves Siemens Solar

1994 Joins NREL as director

1997 Moves to head ASE Americas

2001 *Nonexecutive Chairman SunPower*

2006 *Joins Applied Materials*

2016 *Appointed to lead solar at US DOE*

Charlie Gay must be the best connected person in the photovoltaics industry and one of the most respected and best liked.

Like many others, he joined the sector by mistake. Realizing at an interview for the oil industry that "I had just spent my whole time at college studying something I didn't want to do as a career" [49], he rewrote his resume to focus on materials research, and in 1974 got a job welding solar cells at Spectrolab. Over the next 3 years, he was given free rein to undertake research in several areas, which "provided the groundwork for a career in photovoltaics."

When former Spectrolab colleague Bill Yerkes^{qv} sold his start-up company to Arco^{qv}, he persuaded Gay to join them and head up the research and development activity. Much of the work, at this time, was on establishing a pilot production line, and in particular developing cell processing techniques that were better suited to volume manufacturing. Gay was trying to build his research team at Arco, but found that the best graduates wanted to go into the nuclear industry or better funded big laboratories. After 6 months had failed to deliver a single recruit, he persuaded his brother Bob to join the team.

The combination of Bill Yerkes' inspiration and enthusiasm, Charlie Gay's thoroughness and attention to quality, and Atlantic Richfield's resources made Arco Solar the world's most successful PV company in the 1980s. At the end of the decade, when Atlantic Richfield was selling the company, Gay was appointed president and he went with it to become the first chief executive of Siemens Solar. After leaving in 1993, Gay worked for several months with Chris Eberspacher in a start-up developing thin film solar cells, before being appointed the director of NREL^{qv} in 1994.

Two weeks later, the Republican Party's "Contract with America," led to NREL receiving a 33% budget cut. The new director had to manage a major retrenchment, while retaining the integrity of external research contracts and the morale of his staff. He moved in 1997 to head up the American operations for ASE^{qv}, and was succeeded at NREL by Admiral Truly.

Gay was one of the cofounders of Greenstar Foundation, an organization set up to enable villages in the developing world to use solar power for health, education, and environmental programs and for Internet access through wireless or satellite communications.

Dr. Gay subsequently accepted executive, advisory, and nonexecutive roles with technology enterprises, including SunPower Corporation^{qv} of which he was nonexecutive chairman during 2001–2005 and Applied Materials. In 2016 he was appointed to head up the "SunShot" solar program at the US Department of Energy.

Prof. Adolf Goetzberger

Adolf Goetzberger founded and directed Europe's foremost solar research institute, the Fraunhofer Institute for Solar Energy in Freiburg.

The young Goetzberger's studies were interrupted by the Second World War and its aftermath, but he received a doctorate in physics from the University of Munich. After 3 years with Siemens, he spent 10 years in the United States, first with the coinventor of the transistor, William Shockley, at his laboratories in Palo Alto and then at the birthplace of the solar cell, Bell Laboratories.

Dr. Goetzberger returned to Germany in 1968 to work for the Fraunhofer Institute for Applied Solid State Physics, becoming its director in 1971. His desire to start a solar energy institute at Fraunhofer met resistance, not least from the German government, because solar energy was considered too insignificant.

Goetzberger eventually prevailed and provided the necessary funding for the Fraunhofer Institute for Solar Energy Systems F-ISE to be founded in 1981, initially with 18 staff. Much of the early work was on thermal and passive solar. As director of the Institute he ensured that it always adopted a holistic, whole system approach to solar energy.

Similarly, as the PV work grew over the years, he ensured that it covered not just devices but also storage, power conversion, and other systems aspects.

While directing F-ISE, Dr. Goetzberger remained actively involved in its scientific work, in areas like fluorescent collectors and latterly agrophotovoltaics. He was personally directly involved in the self-sufficient solar house built in Freiburg in 1992, the year before he retired as the director.

Adolf Goetzberger has held advisory and academic roles at universities in Germany and overseas including the Instituto de Energía Solar at Madrid[qv]. He has also been an active supporter of broader organizations, as a director of the International Solar Energy Society[qv] for 12 years, its president from 1991 to 1993, and president of the German equivalent. Among many other awards, he has a virtual "grand slam" of solar prizes, including the Karl Böer Medal[qv], the Becquerel Prize[qv], the William Cherry Award[qv], and the ISES Farrington Daniels Award[qv].

His retirement from the management of F-ISE has not prevented his ongoing involvement in its work, which still continues. Further details of his achievements and those of F-ISE appear in his chapter in the *Solar Power* book [43].

Dr. John Goldsmith

John Goldsmith's career in photovoltaics started in the space sector and dates all the way back to the 1960s.

He completed his graduate studies in physics in Pennsylvania and at the University of California, Los Angeles. Dr. Goldsmith's work in solar research, engineering, and project management started at General Electric's Missile and Space Division in 1960.

Three years later, he moved to the Jet Propulsion Laboratory[qv], where he worked, *inter alia*, on quality standards and test procedures for solar modules. Here he helped the US Coast Guard with their early PV-powered navigational aids. In the 1970s, he was appointed deputy manager of the US Energy Research and Development Administration's Low-Cost Silicon Solar Array Project, the US government's program to help terrestrial PV companies to reduce costs.

In 1978, Goldsmith transferred to Solarex Corporation[qv], initially as head of their space activities but subsequently as vice president in the company's general management. In this role, he briefed President Jimmy Carter on the status and potential for photovoltaics in August 1979.

Goldsmith moved in 1994 to join Pilkington Solar where he was appointed general manager of their international division based in the United States. Here he focused on architectural applications of photovoltaics, based on the developments of Flachglas Solartechnik[qv].

He left to establish his own advisory business Goldsmith and Associates in 1997, and retains an active interest in the sector from his home in Maryland. John has served on the organizing committees for the main PV conferences in both the United States and Europe and chaired the 1978 Washington IEEE conference.

Prof. Michael Grätzel

Michael Grätzel is no semiconductor specialist, yet lends his name to the highest profile new solar cell invention of our time frame.

He earned his doctorate in Natural Sciences at the Technical University of Berlin, and started researching alternative fuel production in the wake of the second oil crisis, first as a postdoc in the United States and then at the Ecole Polytechnique Fédérale de Lausanne[qv].

When the oil price dropped again in the 1980s, funding for alternative fuels became hard to find, but Dr. Grätzel believed his technique for using sunlight to split water into hydrogen and oxygen – introducing colloidal particles as a catalyst to sensitize TiO_2 – might also be applied to photovoltaics. He was proved right when his research on what came to be called dye-sensitized solar cells (see Section 4.6) started to bear fruit in the late 1980s.

The invention attracted great attention when published [67] with coauthor Brian O'Regan in 1991. The following year, Grätzel presented his work at the 11th European PV Conference[qv] in Montreux, receiving a lot of attention, but also, he says, "a lot of scepticism" [123]. Solid-state photovoltaic specialists wanted to know how the cells were tested, how efficiency was calculated, and all sorts of similar details that were entirely new to his group of chemists in Switzerland.

Michael therefore set about building expertise in characterization and measurement. He had cells tested at the Fraunhofer Institute[qv] and NREL[qv]. New methodology needed to be developed to cater for the slower response of the dye-sensitized "Grätzel cells" and appropriate reference standards developed. He found NREL especially helpful, perhaps because they too had worked on photolytic fuel production, and he learned a lot from Keith Emery. As the research progressed, Michael facilitated technology transfer to several companies.

Grätzel continues to direct the Laboratory of Photonics and Interfaces at EPFL[qv]. He has authored several books, more than a thousand publications, and is one of the most highly cited chemists in the world. He has received too many awards to list here, and served as visiting professor to dozen of institutions around the world.

Latterly, dye-sensitized solar cells have achieved laboratory efficiencies of over 10% and are produced by several companies in different parts of the world. The technology also spawned research on perovskites where efficiencies of over 20% have been reported.

Prof. Martin Green

The most distinguished academic in crystalline solar cell research, Martin Green set the standard for solar cell performance for much of our time frame and subsequently.

Green obtained his degree from the University of Queensland and completed his PhD at McMaster in Canada, specializing in solar energy. He started the Solar Photovoltaics Group at the University of New South Wales[qv] in 1974 to work on the development of silicon solar cells.

Martin's group set many solar cell efficiency records in the 1980s and 1990s and was the first to produce a 20% efficient silicon cell in 1985. This success can be attributed to his focus on factors limiting silicon solar cell performance, and innovative approaches to overcoming them, such as oxide passivation, surface texturing, laser-buried grids, and rear diffusion (see Section 4.2).

A touch of good old Australian competitiveness undoubtedly helps too. Green took great pride in winning the World Solar Challenge car race in 1996 with Honda, who had won it 3 years earlier with competitive high-efficiency cells from SunPower[qv]. His instinctive response to having just lost the world record, when we met [124] was simply "we'll get it back."

Although most of his focus has been on monocrystalline cells, Green has worked on polysilicon including a polycrystalline silicon thin film on glass technology. He cofounded Pacific Solar (later CSG Solar) to commercialize the UNSW group's research in 1995, and served as a director for some years.

In a country, where little was done to support PV at the national level, Martin's preeminence is remarkable. He is justified in claiming that his alumni sit in the boards of 70% of the Chinese producers, which latterly dominate the world market.

He has written books [125] and a large number of learned papers. His national and international awards are too numerous to list, including the William Cherry Award[qv], Böer medal[qv], and the Australia Prize; he has latterly been elected to London's Royal Society.

Prof. Yoshihiro Hamakawa

Professor Yoshihiro Hamakawa was a uniquely eminent character, from a culture that traditionally favors the collective over the individual.

Yoshihiro Hamakawa won his doctorate in 1964 and spent much of his career at Osaka University as professor of Electrical Engineering Science.

Among many achievements relevant to the PV sector, probably the most prominent were his pioneering work on multilayer or "tandem" amorphous silicon solar cells, development of modulation spectroscopy for characterization of materials, and valency electron control of amorphous silicon carbide films.

Prof. Hamakawa was proud if not always protective of his students. A conference regular recalled [126] the Q&A session after one such protégé had presented a paper at the Stresa Conference[qv]. As soon as the first question was asked, Hamakawa jumped up from the front row of the audience, grabbed a microphone, and gave a comprehensive answer. This was repeated a couple of times and then a particularly tricky question followed. All eyes turned to the professor, but he stayed seated, gesturing to the hapless presenter on stage.

Hamakawa was instrumental in the Japanese Sunshine Project from the start, serving as a board member since 1973, and advised MITI[qv] and other national agencies. He later moved to Ritsumeikan University where he was appointed Vice Chancellor from 1998 to 2000. He has authored hundreds of papers and dozens of books.

In the early PV era, Yoshihiro was the Japanese PV pioneer most active internationally, and he served on most of the major conference committees. He has received many awards, including the William Cherry Award[qv] in 1994. In 1995, he received a Purple Ribbon from Emperor Heisei. He died in 2016.

Brian Harper

Although never active in the upstream photovoltaics industry, Brian Harper was one of the earliest adopters of the technology and an important champion for solar energy in the Middle East and beyond.

His solar career began as a direct result of the first oil crisis, which slashed demand from the oil companies for the forecourt canopies his sheet metal company produced. Demand for equipment housings in the Gulf region increased, however, and he turned to solar power as a way of cooling them. He applied a similar approach when he branched out into sustainable buildings in off-grid sites in the United Arab Emirates (UAE). During a year in the United States, he worked with Arco Solar[qv] and then returned to the Middle East in the late 1970s to establish Solar Energy Centre (SEC) in Bahrain with his wife Heather.

SEC supplied solar systems and Arco panels throughout the region and went on to add batteries to its range, distributing initially for Exide and later Gould National Batteries (GNB). Brian applied his knowledge of solar applications to support the development of low-maintenance sealed batteries, and SEC later introduced its own range.

Brian was a serial innovator who always had another idea for the promotion of smart energy technology up his sleeve. The huge international attendance at the funeral, following his tragic death in 2014, attests to the affection in which he was held by those who worked with him.

Dr. Peter Helm

Peter Helm gained his PhD in economics and business having obtained a degree in mechanical and processing engineering at the Technical University of Munich. After a spell as scientist at the international Battelle Institute, he joined WIP working on international development projects.

Helm had to look up Managua in an atlas, when he was posted to his first project, working with the German Bank for Reconstruction KfW, on forestry and agriculture in Nicaragua. Other projects in Africa followed, before a chance meeting with the EC's Wolfgang Palz[qv] led to WIP being appointed to monitor the EC's pilot projects (see Section 13.1) in the early 1980s.

Dr. Helm has since served for many years as WIP's scientific director and led its involvement in a wide range of PV-related advisory activities and in particular the management of the European PV Conferences[qv]. He acts as contracted expert providing scientific and project monitoring assistance to the European Commission and other international and German organizations.

Prof. Robert (Bob) Hill

Bob Hill eschewed the normal inclination of British academics to maintain a respectful distance between academia and industry. He pursued a very holistic approach to research in the PV sector.

He obtained his doctorate in solid-state physics from Imperial College, London where he had gained his degree, and joined the physics department of (what is now) Northumbria University[qv] in 1971, after spells in the nuclear industry and Newcastle University.

Hill soon developed a passion for PV and by 1978 was working on satellite solar panels. Ever the pragmatist, he realized that new universities could compete better by applied, rather than fundamental, research; so he concentrated on practical issues like solar cell testing. Shortly after he was appointed Professor of Opto-Electronics in 1984, Hill established NPAC[qv] (the Newcastle Photovoltaics Applications Centre), which he directed until his retirement.

His personal academic output was large and influential: 12 books, including one for the UN [127], learned papers, and reports. He served on many technical and advisory committees and chaired the 12th European PV Conference^{qv} in 1994. He was a founder director of the British Photovoltaic Association PV-UK and served as its chair in 1994–1995 and 1999.

Prof. Hill will be best remembered for the breadth of his insights and interests. He placed technology alongside social, economic, and environmental aspects. He was not an ivory tower academic, but could explain scientific concepts cogently to any audience. He also inspired those who worked with him, including his successor Nicky Pearsall^{qv}; it is no surprise that many of them have gone on to distinguished careers in the sector.

Bob Hill's early death at just 62 in November 1999 was a huge loss to the photovoltaics community. A prize in his name for promoting PV applications in the developing world was first awarded in 2005. His son Ben followed him into PV, working at leading companies in the sector. Bob will be remembered for his unfailing cheerfulness and optimism about the future of PV and the sheer breadth of his knowledge and insight across a huge spectrum of interests.

Dr. Winfried Hoffmann

Summarized timeline

1979 Joins Nukem

1985 Appointed head of solar at Nukem

1986 Proposes JV with MBB solar division

1994 Leads JV with DASA to form ASE

1994 ASE acquires Mobil Solar

1997 First serves as EPIA president

2001 ASE becomes RWE Schott Solar

2001 Appointed to SMA board

2007 Joins Applied Materials

Europe's best connected solar industrialist, Winfried Hoffmann was key to the formation and management of Germany's most prominent early PV companies.

Having completed a doctorate in biophysics, Dr. Hoffmann shed his original plan to go into teaching, joining instead RWE's nuclear joint venture Nukem^{qv} at Hanau. He soon transferred to the recently formed thin film solar cells research group, working in the research department on cadmium sulfide/copper sulfide (Cu_2S/CdS) cells in collaboration with the Stuttgart^{qv} group of Werner Bloss^{qv}. Six years later in 1985, he took over the leadership of the solar group.

Although the CdS work was achieving reasonable results (around 6% efficiency), amorphous silicon was by then the global thin film of choice. Hoffmann therefore explored a possible merger between Nukem's solar business and that of MBB, which was working on amorphous silicon near Munich. The outline of an agreement was ready in 1986, just before the Chernobyl disaster. Nukem were then unwilling for all employees of their environment-friendly solar business to leave Hanau wholesale and move to Munich, so the deal fell through – for the time being.

As one door closes, another opens. Two days later, Nukem received a contract for a joint research program on MIS (metal–insulator–semiconductor) technology. In due course, a pilot plant was set up at about 2 MW capacity, but scaling up looked challenging.

In 1994, Dr. Hoffmann led the merger of Nukem's solar division with that of Deutsche Aerospace (DASA), which now included the PV businesses not just of AEG[qv] but also of MBB; so the merger planned 8 years before happened in the end. Hoffmann was appointed Managing Director of this new joint venture called ASE[qv]. He had great faith in ribbon processes for crystalline silicon, so ASE acquired Mobil Solar the same year, renaming it ASE America. He recruited Charlie Gay[qv] as president of the US company 3 years later.

Subsequently, DASA withdrew from the ASE joint venture, and some months later Schott invested in what then became RWE Schott Solar. Hoffmann remained chief executive until Schott bought out RWE in 2005. Charlie Gay had the opportunity to return Winfried's favor in 2007, enticing him to become Chief Technology Officer at Applied Materials (which Gay had joined the previous year). Dr. Hoffmann has also served on the advisory boards of SMA[qv], Fraunhofer ISE[qv], and Solar-Fabrik.

Hoffmann has been heavily involved in representative associations, serving on the board of Germany's BSW and EPIA[qv], where he was a long-standing president during and after our time frame. He has written a book [128] about the PV sector, and won many awards, including the Bonda[qv] and Becquerel[qv] Prizes. He now acts as an independent consultant.

Theresa (Terry) Jester

Theresa Jester's claim to fame goes well beyond the fact that she was one of the first female engineers in the solar industry. She joined Arco Solar[qv] in 1979, fresh from her engineering BSc at California State University at Northridge. She started working in Bill Yerkes' "skunkworks" on innovative products such as solar refrigerators.

Jester progressed to become vice president of engineering, responsible across the full production operation, from ingot growth, wafer production, and cell processing to module assembly and product development. In this role, she became a leading expert on safety in PV manufacturing, and was subsequently appointed to the Advisory Board of the National Center for Photovoltaics.

She stayed with Arco Solar through its changes of ownership first to Siemens Solar[qv], subsequently, Shell Solar, where she became vice president of operations, and eventually Solarworld.

Latterly, Terry Jester has held positions with SunPower Corporation[qv], Hudson Clean Energy – rejoicing with the title of "Entrepreneur in Residence" – and Calisolar, before being appointed Chief Executive of Silicor.

Bob Johnson

Bob Johnson is one of the world's most knowledgeable experts on the development of the terrestrial PV industry. Before discovering solar power, he was an engineer, first in the US Navy and then in the power industry, spending time on nuclear energy for the infamous Admiral Rickover, and later in vacuum systems.

In 1974, a previous boss of Johnson's, Steve Dizio, then president of Shell's subsidiary SES[qv], persuaded him to join them. Bob was appointed as Sales and Marketing Manager, despite his technical background – and the fact that company had, at the time, little to sell. In this role, he was also responsible for applications engineering.

In the early 1980s, when it was becoming clear that SES faced insurmountable problems in turning its technology into a viable commercial product, John Day[qv] approached him to join Strategies Unlimited[qv]. Bob did so and soon made the company's PV work his own, allowing John to focus on optoelectronics. Undoubtedly, his background led to Bob's focus on evidence-based research; he went to great lengths to maximize the industry's input to the reports, which the company produced. He demanded the same attention to details from his team; Paula Mints recalls [76] daily 4 h lectures on the minutiae of PV for the first 2 weeks after she joined Strategies Unlimited.

Bob was very sensitive to the dynamics of the sector. Though most of his customers were multinational companies, he maintained close and friendly links with, and was trusted by, many of the smaller independent companies too. He seldom missed a major international PV conference.

Bob Johnson retired from Strategies Unlimited in 2001 when it was sold.

Dr. Lawrence (Larry) Kazmerski (Kaz)

Larry Kazmerski (often referred to simply as Kaz) is one of the most widely recognized research engineers in the sector.

Dr. Kazmerski achieved his PhD in electrical engineering from the University of Notre Dame, and went on to the University of Maine, where he worked on thin film PV as part of the team, which reported the first copper–indium diselenide (CIS) solar cells [61].

He moved to Golden Colorado in 1977 to become the first PV staff member at the Solar Energy Research Institute, the forerunner of NREL[qv]. On a visit by President Jimmy Carter, the following year, Kazmerski's quick reactions are credited [82] with saving the commander-in-chief from a possible early death by electrocution.

Dr. Kazmerski has become a leading world authority on semiconductor materials and devices, thin films, surface and interface analysis, scanning probe microscopy, nanoscale technology, high-temperature superconductivity, and semiconductor defects. He has hundreds of papers to his name, and has received awards for the development of methods and instruments to characterize the electro-optical, chemical, and structural properties of materials at atomic- and microscales.

He has devoted substantial efforts to the dissemination of PV history and technology through video and exhibits. He has written four books, and served as editor-in-chief of *Renewable and Sustainable Energy Reviews*. He has also held adjunct professorships at the University of Colorado, Colorado School of Mines, and the University of Denver and served as a director at Suntricity Corporation.

When the US Department of Energy created the National Center for Photovoltaics (NCPV), based at NREL in 1996, Kazmerski was appointed to lead its measurements and characterization division. He was appointed overall director of the NCPV two years later.

His contributions to the sector have won many awards, including the William Cherry Award[qv] and the Karl Böer Medal[qv]. He has also been elected a member of the National Academy of Engineering, reserved for those who have made outstanding contributions to engineering research, practice, or education and to pioneers of new and developing fields of technology.

Since retiring from NREL, Kaz has continued to travel extensively around the world passing on the benefits of his knowledge and expertise.

Dr. Zoltan Kiss

Serial entrepreneur Zoltan Kiss was the founder of Chronar Corporation, for a time one of the most prominent amorphous silicon developers and producers.

Kiss (pronounced "Kish") was born and raised in Hungary. In 1950, in his teens, he escaped the Soviet-dominated country to Austria, where he finished his schooling. He went on to Toronto University, gaining a PhD in Physics in 1959 and a post-doctorate at Oxford University.

Becoming restless after 9 years in his first job at RCA in Princeton, he left to set up Optel in 1970, taking a few RCA colleagues with him. Optel introduced an early liquid crystal watch, but proved unable to meet demand and was soon overtaken by other producers, notably from Japan.

Dr. Kiss had moved onto his next start-up Chronar Corporation^{qv} by the time Optel collapsed. Chronar was a producer of amorphous silicon panels and products. In its heyday, it had factories not only at its headquarters in New Jersey but also in four overseas countries, which together at their peak were among the world's largest thin film solar cell producers.

Following his departure from Chronar in 1989, Dr. Kiss established EPV (Energy Photovoltaics) to focus on amorphous silicon manufacturing plant and other thin film technologies, employing several ex-Chronar staff.

Latterly, James Groelinger arranged a buyout of EPV, replacing Kiss as CEO. Zoltan has since established various further companies, including Amelio Solar, Nanergy, and Renewable Energy Sources in New Jersey and Solar Thin Films in Hungary. Together with his architect son Gregory, he has also been involved with building-integrated photovoltaics through Terra Solar.

Zoltan Kiss was described by former colleague Jonathan Allen [129] as "an adventurer in the business sense, willing to try things that offer the possibility of doing some real good, even though there are substantial business risks involved."

Dr. Yukinori Kuwano

Yukinori Kuwano is a world expert in amorphous silicon and heterojunction technology, who progressed to the presidency of a global electronics company (Fig. 10.3).

Kuwano joined Sanyo^{qv} in 1963 after his bachelor's degree at Kumamoto University. From the late 1970s to mid-1980s, he managed Sanyo's research center in Hirakata, leading the work on amorphous silicon solar cells and

Figure 10.3 Yukinori Kuwano [130] celebrates the 20th birthday of his rooftop solar system.

writing several books on the subject. He completed a doctorate at Osaka University in 1981. His group developed and manufactured the world's first mass-produced integrated amorphous silicon solar cell module.

Kuwano is credited as one of the first to identify the superior low light performance of amorphous silicon cells, leading to their adoption in indoor solar products. His group also invented the silicon heterojunction (HIT) solar cell.

At the fourth Asia-Pacific PVSEC[qv] Conference in 1989, he introduced the "Genesis Project," a proposed Global Energy Network Equipped with Solar Cells and International Superconductor Grids. Dr. Kuwano has received several awards for his achievements including the PVSEC Award[qv].

In the late 1980s, Dr. Kuwano moved to Sanyo North America, returning as general manager of all Sanyo's research and development in 1993. He was appointed managing director of Sanyo Electric three years later and president and chief executive in 2000, replacing Sadao Kondo who, paradoxically, had resigned following a scandal about underperforming PV systems.

Since his subsequent retirement from Sanyo in 2005, he has held advisory and nonexecutive positions with Sanyo and other companies, including Daiwa House, Tabuchi, and Optex. He has never lost his enthusiasm for solar energy – becoming president of the Photovoltaic Power Generation Technology Research Association – and he has maintained contact with friends in the sector.

Figure 10.4 Peter Lecomber (standing) with Walter Spear[qv] (seated) [25].

Dr. Peter LeComber

A crucial coworker of Professor Spear[qv], Peter LeComber (Fig. 10.4) shares the credit for much of his pioneering work on amorphous silicon outlined below.

He met Walter Spear while lecturing at Leicester University, and together they went to Dundee University in 1969 to establish the Carnegie Laboratory of Physics working on noncrystalline solids.

Spear and LeComber coauthored the paper [131] that first announced the semiconductor properties of hydrogenated amorphous silicon in 1976. Together, their work on thin film semiconductors was crucial not just to the PV industry but also for the development of electronic displays, flat-screen televisions, and other large-area devices.

LeComber was appointed professor of solid-state physics at Dundee in 1986, a chair created specifically for him. Shortly after being elected a Fellow of the Royal Society in 1992, he died of a heart attack ending a brilliant career and depriving Walter Spear of an irreplaceable partner.

Roger Little

Roger Little is a master of endurance both within the PV sector and beyond.

He obtained his degree in physics from Colgate University and a masters in geophysics from Massachusetts Institute of Technology (MIT), before going on to found Spire Corporation[qv] in 1969.

Little oversaw the development of the company from a producer of space cells into the terrestrial PV business and then to a supplier of production equipment and turnkey plants. It was his strategic approach that enabled Spire to find profitable market niches, avoiding direct competition with larger, often loss-making, companies.

Even as the company grew, Roger remained personally the public face of Spire Corporation. Because the company sold extensively within the PV sector, it regularly exhibited and presented at exhibitions and conferences. Participants who were up early would often see Roger pounding the streets. He was an accomplished triathlete who, not content with the simple Olympic version of the sport, completed a quarter century of "Ironman" events – 2.4-mile swim, 112-mile cycle ride, and 26.2-mile marathon run.

He served on many committees and advisory boards related to the PV sector (including as president of SEIA[qv]), small business, innovative research, technology transfer, and the development of renewable energy policy.

Roger Little served for 44 years as chief executive officer of Spire from its formation until he subsequently retired at the end of 2013, handing over to Rodger LaFavre, while remaining nonexecutive Chairman.

Dr. Joseph Lindmayer

Among the scientist-entrepreneurs of the early PV industry, Joseph Lindmayer was rare for his intellect, and even rarer for the prosperity he achieved (Fig. 10.5).

Figure 10.5 Dr. Lindmayer at European Conference [25].

Educated at the Technical University of Budapest, Joseph made a perilous escape from the land of his birth as the Soviet tanks rolled in to crush the 1956 Hungarian Uprising. After a master's degree in the United States, he gained his PhD from the University of Aachen.

Returning to the States, Joseph (he personally never used the handle "Joe" given to him by American colleagues) became head of physics research at Sprague Electric. With colleague Chuck Wrigley, he wrote *the* book [132] on solid-state electronics in 1965 at the age of 33. He also served as a visiting physics professor at Yale University.

In 1968, Dr. Lindmayer moved to Washington to become Head of Physics at Comsat, the Communications Satellite Corporation. He apparently showed little initial interest in solar cells, but eventually took up the challenge of increasing the performance of silicon space cells. Within 3 years, Comsat's cell efficiencies had risen from 10.5 to 13% [43]. While working at Comsat, he met fellow countryman, Peter Varadi[qv], then head of chemistry. They started discussing the terrestrial potential of photovoltaics, and during their 1972 New Year's Eve party agreed to set up in business together. They would take some months to raise capital and then leave Comsat. Lindmayer insisted the new company name must end with an X – and so Solarex[qv] was born.

Despite its low initial capital base, the company made good progress into the market, thanks in part to Lindmayer and Varadi's ability to come up with ideas for cost reductions. The company was world market leader, when Amoco invested in 1979, and the price they paid made a good return for the founders and angel investors.

Lindmayer always retained his thick accent (somewhat reminiscent of Henry Kissinger), and was not a great extrovert. But he was intellectually persuasive and had a good eye for the grand gesture. He had to be the winning bidder at the Cannes auction (see page 88); he created Solarex' low-cost silicon business Semix and named their solar-powered solar module factory the Solar Breeder, as a jibe to atomic power. "A nuclear breeder eventually runs out of fuel," he said. "That's not a breeder. A true breeder is one that doesn't run out."

In 1983, Amoco acquired the rest of Solarex, at a much less generous price. Lindmayer left and (retaining his predilection for names ending X) set up Quantex Corporation and its sister companies Optex and Photonex. Prevented by his noncompete agreement with Amoco from working in PV, these companies operated in the related field of photonics; for detectors, X-ray replacement and other applications. He also served as a nonexecutive director of Intersolar Group[qv].

Lindmayer died in 1995. Though he never saw the PV industry approaching maturity, his contemporaries all acknowledge the major contribution he made to the early days of terrestrial solar power.

Prof. Antonio Luque Lopez

Antonio Luque is one of Europe's most distinctive and distinguished photovoltaics academics.

He obtained degrees from the Technical University of Madrid (UPM) and the University of Toulouse before gaining his PhD at UPM in 1967, while constructing Spain's first laser. In due course, Luque became Professor of Physical Electronics at UPM, working mainly on photovoltaics, with a particular focus on high-efficiency approaches. He was the first to make a bifacial solar cell in 1976. Prof. Luque founded the Institute of Solar Energy[qv] at UPM in 1979, with a major objective being to reduce the cost of solar photovoltaic energy.

Two years later, he started the company Isofotón[qv] in Málaga, and was its chairman until 1989, while retaining his "day job" at UPM. He has also worked extensively on solar concentration and invented the intermediate band solar cell in 1997. He has written a number of books [133], hundreds of articles [134], and has dozens of patents.

Antonio Luque has established widespread contacts around the world, notably in Russia and Belarus, and served on advisory boards in Germany, France, and the United States. He has received many awards, including the Becquerel Prize[qv], William Cherry prize[qv], and Karl Böer Medal[qv].

Prof. Luque continues his academic work, latterly coordinating the "Full-spectrum" project on more efficient use of the full solar spectrum, research and development on intermediate band cells at the Ioffe Institute, and the joint EU–Japan NGCPV initiative on solar energy.

His achievements and views on the PV sector are available in his chapter in the *Solar Power* book [43].

Dr. Paul Maycock

For a man who managed the world's largest government PV support program, Paul Maycock, with his booming voice and forthright views, couldn't seem less like a civil servant.

Maycock's first encounter with photovoltaics was at RCA[qv], in the interlude between his degree and joining the Office of Naval Research. On coming out of the US Navy, he went to Texas Instruments developing scientific calculators and "speak and spell" devices. In 1975, he joined the US Department of Energy[qv] as director of the Photovoltaic Energy Systems Division.

During the Carter presidency, the DOE support program was very comprehensive, as outlined in Section 8.3, and the budget, which had been a few millions per annum, rose substantially. In this bright period, Maycock focused persistently on supporting cost reductions in industry. He also contributed to the important Public Utility Regulatory Policies Act, which was enacted in 1978. Things changed when the newly elected Reagan administration cut the 1981–1982 budget back from the original $150 millions to $56 millions, and Dr. Maycock decided to leave.

He then wrote his book [135] with Ed Stirewalt and founded PV Energy Systems[qv], an information provider in the sector. In this role, Maycock published monthly newsletters and annual market reports, and also undertook specific consultancy projects for the World Bank and others. He also served for several years as nonexecutive Chairman of Solar Electric Light Fund[qv].

He sold his advisory business on retiring early in the new millennium.

Bernard McNelis

Bernard McNelis is long-serving stalwart of the global renewable energy sector, and in many respect its "social secretary." He was a founder of IT Power, which he directed for more than 35 years.

In his early career as a chemist with Chloride Batteries, McNelis was asked to study the embryonic renewable energy sector and assess its relevance for the battery industry. Chloride concluded that they would not get directly involved in the renewable energy business, but by then he was hooked.

Solar Power Corporation Chief Executive Bob Willis[qv], who had been introduced by their UK representative Clive Capps, gave McNelis a job, and he represented SPC at the early European conference in Hamburg (see Section 14.2).

When Exxon reduced their support for SPC, McNelis found himself out of a job, and undertook part-time roles with Solar Energy Developments and General Technology Systems. Having met Intermediate Technology Development Group through ISES[qv], McNelis became involved in a Halcrow-led global water-pumping project for the World Bank, working with Anthony Derrick, Peter Fraenkel, and Fred Treble[qv].

In 1980–1981, McNelis, Derrick, and Fraenkel established IT Power[qv]. PV was mainly used at that time in stand-alone off-grid commercial applications. McNelis, however, realized that its simplicity, reliability, and ubiquity also make it suitable for more personal and social applications, especially in remote communities. This focus soon made IT Power one of the "go to" experts on the deployment of renewables for water pumping, medicine, and rural electrification.

Bernard devoted much time and effort beyond the parochial concerns of his own business. He served on committees for standards, conferences, and the IEA. He became an inaugural member of the UK chapter of the International Solar Energy Society ISES[qv], led by Mary Archer, and has served as its vice president and on the international board. He was instrumental in establishing the UK's first solar power trade association PV-UK. Following the death of EPIA Secretary General John Bonda[qv], its board asked McNelis to hold the fort part-time. He kept things ticking over for almost a year, handing EPIA's affairs over to the next secretary, Murray Cameron, in 2000.

But the characteristic that first comes to mind for most PV old-timers is that he is such good company. He went to all the conferences, and was never seen without a joke on his lips and his camera around his neck. Attendees soon learnt that, after a long day at the conference, you just had to fall in behind Bernard for a few happy hours in the nearest good taverna or bar. Fortunately, he is also an incurable hoarder; this book is greatly indebted to his archive and photographs.

McNelis continued working primarily on rural electrification, based in the United Kingdom or China, until he retired – if he has yet done so!

Dr. Heinz Ossenbrink

A leading expert on PV testing, characterization, and standards, Heinz Ossenbrink received his PhD in nuclear physics from Hahn Meitner Institute, Berlin and joined the CEC's Joint Research Centre[qv] at Ispra in 1982.

He worked on pilot projects (see Section 13.2), visiting and assessing many of the systems. He was involved in the development of test methodology and standards for PV devices and systems, and has many publications on this topic and related fields.

He has served on international standards committees and advised Peter Varadi[qv] on the establishment of PV-GAP.

In 1995, Dr. Ossenbrink was appointed head of the renewable energy unit at Ispra, and expanded research and support activities to energy efficiency, bioenergy, and biofuels. At the same time, he began long service as program chair for the series of European Photovoltaic Solar Energy Conferences[qv].

Latterly, Heinz took over chairmanship of IEC's Technical Committee 82 on international standards for PV systems (from his predecessor NREL's Dick DeBlasio), serving for 14 years until he retired in 2016.

He still sails his yacht on Lake Maggiore.

Stanford (Stan) and Iris Ovshinsky

Iris and Stan Ovshinsky were among the most creative researchers in the early sustainable energy sector, and Stan was one of its great showmen.

Stan Ovshinsky, whose Jewish Lithuanian parents had immigrated into the United States, skipped higher education in favor of training as a toolmaker and starting his own business. Having sold his first company in the 1950s, Stan worked with his brother Herbert to build an electromechanical model of a neural cell and, in doing so, discovered that amorphous materials could perform like crystalline semiconductors. He named the technology "Ovonics" to differentiate it from traditional crystalline electronics.

Around this time, Ovshinsky met his second wife Iris, who made up for his lack of credentials with a string of degrees; she became his business and life partner, when they married in 1959. They founded Energy Conversion Devices[qv] the next year and took it public in 1964, to raise development capital for their amorphous silicon devices.

Stan Ovshinsky was not only an intuitive inventor, but also a natural salesman, not afraid to promote himself or his ideas. According to *Forbes* magazine [136], Ovshinsky claimed that his amorphous devices "'would compete with silicon crystalline transistors,' and in the space of a day ECD stock shot from $57 to $150. The [amorphous] transistor never made much money and the company's stock never returned to those heights, but Ovshinsky did manage to get the word Ovonics (and his name) into the dictionary."

Ovshinsky was also adept at attracting government research grants and instrumental in persuading a succession of deep-pocketed corporations, to fund developments for which the Ovshinskys contributed the ideas. Some of these led to products – most notably amorphous solar cells and nickel–metal hydride batteries – that ECD sought to commercialize through a network of subsidiaries, few of which ever made money.

Stanford Ovshinsky was honored as a "Hero for the Planet" by *Time Magazine* in 1999, and inducted into the US-based Solar Energy Hall of Fame in 2005. Following his death in 2012, a few years after Iris, Forbes offered a kinder perspective: "While Ovshinsky raised – and lost – hundreds of millions of dollars, his contributions to understanding energy management had a lasting impact" [137].

Dr. Wolfgang Palz

The European renewable energy sector owes its prominence in no small measure to Wolfgang Palz.

Palz graduated from Karlsruhe University studying physics and nuclear energy; he then moved to the semiconductor group to work on cadmium sulfide monocrystals, pulled in the university's own laboratory. His doctoral thesis was based on the unexpected response to infrared light he discovered; this coincidentally proved relevant years later when at the European Commission (CEC) he supported a project by Antonio Luque[qv] on the broad-spectrum response of solar cells.

A lifelong Francophile, Palz then relocated to France, initially as an exchange professor at Nancy before moving to Paris to work for 10 years at the space agency CNES as their solar expert. Discovering that Unesco's 1973 solar energy congress in Paris had no plans to cover photovoltaics, Palz was invited to organize a small PV section. He was then sent as one of France's representatives to an inaugural European solar energy meeting in Milan. When the breakout sessions began, only he and CEC's Karlheinz Krebs were in the PV workshop – everyone else had gone to the solar thermal session. Fortuitously, the European Commission's Director-General for Research Günter Schuster took pity and joined them.

A couple of years later when nuclear power became the priority in France, it was suggested that the CEC should pick up renewables research, and Palz was offered a position in Schuster's directorate to head up solar energy under Albert Strub. Over the next 20 years, he presided over successive programs, which did much to develop the embryonic European PV industry, as further described in Section 8.4, including the pilot projects detailed in Section 13.2. He also took the lead in introducing the European Conference[qv] series and the Becquerel Prize[qv], among other initiatives. Astute, if not always diplomatic, Palz was able to support the sector even when it was not a high political priority.

He transferred to the development directorate 4 years before retiring from the CEC in 2002. Subsequently, he worked on an innovative project to introduce hybrid solar- and wind-powered systems for electricity supply and Internet communications in village schools in Central and South America. He has also served as Chairman of the World Council for Renewable Energy.

Dr. Palz has received several awards, including Germany's Order of Merit and the Becquerel[qv] and Mouchot[qv] Prizes. He wrote his first book on solar energy [138] for Unesco in 1978 and has since written and edited a series of books on renewable energy, including one on solar power [43], where you can read further details of the PV sector and his many achievements therein.

Figure 10.6 Nicola Pearsall with Bob Hill en route to Freiburg in 1989 [68].

Dr. Nicola (Nicky) Pearsall

An influential researcher in her own right, Nicky Pearsall is inevitably often thought of as half of a double act with Bob Hill[qv] (Fig. 10.6).

Having graduated in physics from Manchester's UMIST, she moved to do her PhD at Cranfield (which she liked) working on electrolyzers (which she didn't like).

So she transferred to Dr. Tim Coutts' PV team, working initially on cadmium sulfide/copper sulfide space cells for the Royal Aircraft Establishment.

When Coutts later moved to Bob Hill's optoelectronics department at University of Northumbria[qv], she was offered a research post and moved to complete her PhD there. The team decided to steer clear of the more competitive and fashionable research areas in crystalline and amorphous silicon, specializing instead on III–V compounds such as gallium arsenide and indium phosphide.

Pearsall also shared Hill's view about the importance of systems research and happily took on the day-to-day management of the Newcastle Photovoltaics Applications Centre[qv] as soon as it was established. On Bob's death in 1999, she was the obvious "go-to" successor and was appointed professor soon afterward.

Dr. Pearsall has an impressive body of academic work and has served as Vice President of Eurec and on many academic and conference committees.

Dr. Morton (Mort) Prince

Mort Prince is an undisputed PV pioneer. He was there at the very start, and stayed in the sector until he retired.

After gaining his doctorate at Massachusetts Institute of Technology, Dr. Prince was hired by William Shockley to work on early transistor technology at Bell Laboratories, initially on minority-carrier properties. In late 1953, Prince joined Gerald Pearson to work on device characterization and applications. When Pearson had produced the first silicon solar cells with Daryl Chapin and Calvin Fuller, he called Prince over to witness their tests. Years later, he recalled this momentous occasion without embellishment, "the ammeter swung to the right."

The first cells had an efficiency of about 6% (which early prototypes still delivered years later, according to NREL[qv]). Working with Ed Stansbury, Prince progressively increased efficiencies toward 9%. He published a seminal publication in 1955 about the performance potential of the Si solar cell [139].

Dr. Prince left Bell Labs in 1956 to become director of R&D at Hoffman Electronics, leading the efforts to establish PV as the preferred option for power in space. With Gene Ralph, he was responsible for the first solar-powered satellite, Vanguard I, in 1958.

He was also responsible for the solar panels used for the first solar transcontinental broadcast. The caption to Fig. 10.7 reads:

> Dr. Morton B. Prince (left) and Lt. Col. Allan T. Burke check the two solar energy converters which made possible the first transcontinental radio conversations powered entirely by the sun. The 20-square-foot panel in the foreground was installed at Fort Monmouth, New Jersey, and the larger panel [not shown here] is atop Hoffman's Semiconductor Center in El Monte California. Each panel contains 7,800 solar cells.

Dr. Morton B. Prince (left) and Lt. Col. Allen Y. Burke check the two solar energy converters which made possible the first transcontinental radio conversations powered entirely by the sun. The 20-square-foot panel in the foreground was installed at Fort Monmouth, N. J., and the larger rectangular panel in the background is atop Hoffman's Semiconductor Center in El Monte, Calif. Each panel contains 7,800 solar cells.

Figure 10.7 Mort Prince and Allan Burke [140] with Hoffman panels for transcontinental broadcast.

He joined ERDA, the Energy Research and Development Administration, when it was founded in 1975, and stayed when it was merged 2 years later into the US Department of Energy^{qv}, where he joined Paul Maycock^{qv} as deputy director of the Photovoltaic Energy Systems Division.

His contributions to the sector over a long career have been recognized by many awards, including the Becquerel Prize^{qv}.

Dr. Paul Rappaport

Paul Rappaport was one of a select band of early PV researchers and the founding director of the US' leading national research center; so he merits a profile here, even though he died just a few years into our time frame (Fig. 10.8).

Figure 10.8 Paul Rappaport greets President Carter at SERI in 1978 [25].

Paul Rappaport earned degrees from Carnegie Institute of Technology, and his doctorate at Arizona State University. He joined the Princeton research laboratories of the RCA Corporation in 1949, working initially on transistors and integrated circuits.

At William Cherry's instigation, he started working on energy conversion and solar photovoltaic cells in the late 1950s. Rappaport became a leader in the field of photovoltaics. He was named in patents, wrote articles on the subject, and edited *The Journal of Energy Conversion*.

When in July 1977 President Jimmy Carter's administration created the Solar Energy Research Institute^{qv} (SERI) to work on renewable technologies for the US Department of Energy, Rappaport became its founding director. He brought several RCA researchers with him, establishing the institute in Boulder, Colorado as a major center for solar energy research, before his death in 1980.

Rappaport received the William R. Cherry Award^{qv} in recognition of his work in the photovoltaics field. A national fellowship, the annual Paul

Rappaport Renewable Energy and Energy Efficiency Award, was named in his honor.

Dr. Markus Real

The prominent position achieved by Switzerland in the early residential PV market owes much to Markus Real's enthusiastic campaigning.

Real studied engineering at ETH Zurich followed by a doctorate under Peter Baccini, before joining the Federal Institute of Reactor Research (EIR). Here he represented the burgeoning Swiss industrial interest in solar thermal power plant. In the late 1970s, Dr. Real developed solar measurement apparatus leading to contracts for EIR for monitoring early solar power tower plants. He moved on to PV, installing Europe's first grid-connected system at EIR in April 1981.

Markus Real left EIR to found Alpha Real in 1982 working initially on wind and solar power. In 1985, the company collaborated with Mercedes Benz on a solar-powered car that won the first Tour de Sol (see Section 3.8). Despite the enthusiastic reception to this triumph, he says [40], the automotive experts concluded with certainty that electric cars would never become mainstream.

Two years later, Alpha Real connected Switzerland's first grid-connected PV house, and later that summer launched "Project Megawatt" aimed at installing 333 solar home systems of 3 kW each without subsidies. Thanks to enthusiastic uptake by Swiss citizens, the target was soon achieved. The program also led to the successful introduction of a "net metering" approach whereby householders were paid for any exported power at the same price as the electricity they bought – an improvement on his first grid-connected project, where the system owner was paid in postage stamps for power exports.

Dr. Real stopped developing PV projects in Switzerland in the mid-1990s, and in 1997 joined Peter Varadi[qv] to work on the Photovoltaic Global Approval Program (PV-GAP, see Section 8.8). He has subsequently been involved with other international certification activities and served for many years as chairman of the IEC's photovoltaic systems working group.

His book [141] describes his pioneering work in grid-connected solar systems.

Claude Remy

Archetypal Frenchman Claude Remy built his country's largest PV company, sometimes thanks to – but often despite – the national, political, and industrial support he received. Remy spent 18 years in the army, which had supported him through college, before joining CGE subsidiary Alsthom, where he progressed to group director, before being invited to take over as chief executive at Photowatt^{qv}, the recently established subsidiary of CGE's SAFT battery company in 1981.

He presided over a period of growth that established Photowatt as Europe's largest PV producer, with support from the French government and the European Commission, and continuing backing from SAFT and its coinvestor Elf. In the run-up to the privatization of CGE, it became more commercial, telling SAFT boss Georges Chazot, in Remy's presence, to "cease this adventure" with Photowatt [142]. The sale of a majority stake to Chronar France was concluded in July 1987.

Colleagues Robert de Franclieu and Emmanuel Fabre joined Remy in a visit to Zoltan Kiss at Chronar Corporation^{qv}, but they evidently failed to persuade him that his thin film cells were unsuitable for most of Photowatt's markets. Unhappy with the outcome, Remy was dismissed, but having retained his employment contract with SAFT, he returned there. De Franclieu joined him a few days later.

Elements within the French government were unhappy too, suggesting that it might be possible to exploit an irregularity, that Chronar had not obtained the prior authorization required for non-European entities buying significant parts of French companies. Remy and de Franclieu were encouraged to find a way of bringing Photowatt back into French hands.

This proved to be a long and daunting quest [143], as the previous parent companies did not want to retake control, so a new enterprise Solar France was formed with capital from Remy, de Franclieu, other Photowatt staff and well-wishers, and loans from others. They managed to secure the six million French francs (just under €1 m) hours before the deadline in March 1988, and Solar France bought back Chronar's stake, with Remy reappointed chief executive. This was not the end of the turmoil; eventual stability for the company needed three further "strokes of luck" [142], involving strong support from the European Commission's Wolfgang Palz^{qv}; 50 ingots of gold in Auchan supermarket plastic bags; and a strategic investment by Shell^{qv}.

It was Shell's later exit that led to the acquisition of Photowatt by the Canadian ATS group in mid-1998. Remy retired the following year, subsequently establishing a research enterprise to develop a novel fast silicon ribbon process using sacrificial carbon substrates.

Figure 10.9 Claude Remy receives first Bonda Prize [68] with his wife Olena.

Claude Remy was the first recipient of the Bonda Prize^{qv} in 2000 (Fig. 10.9).

Alain Ricaud

Alain Ricaud obtained his engineering degree at the Ecole Supérieure d'Electricité, and his MBA in Paris. He started his industrial career at IBM, before spending 2 years teaching physics in Algeria.

He was appointed Managing Director of France-Photon[qv], a joint venture of Leroy-Somer and Solarex[qv] in 1979. The company operated with some success in international markets, but received little national support in France and was later merged into Photowatt[qv], so Ricaud transferred to Solarex in the United States.

In his 3 years as general manager of Solarex' crystalline division, he was responsible for the transfer of manufacturing to the "solar breeder" plant in Frederick Maryland.

In 1989, Ricaud returned to France and became Managing Director of Solems, the French part of a joint venture between Total and MBB groups, developing amorphous silicon thin films for photovoltaic and electro-optical applications with Phototronics in Germany.

After Solems closed, Alain Ricaud cofounded Cythelia Consultants in 1994, specializing in new energy technologies and positive energy buildings. He also served for a year as a director of HCT Shaping Systems[qv].

Latterly, he has written books, papers [143], and newsletters and taught at various institutions including the École Polytechnique Fédérale de Lausanne[qv].

Dr. Hermann Scheer

Although his principal role was not primarily within the PV sector, politician Dr. Hermann Scheer merits mention because of his influence over the growth of the wider German and European renewables industries.

Scheer was a tireless campaigner for renewable energy ever since his involvement in the student movement in the late 1960s, when he also became a member of Germany's Social Democratic party (SPD). He was elected to the Bundestag as an MP in 1980.

Scheer played a key role in introducing the country's solar-roof programs and the renewable energy law, as further described in Section 8.4; he moved to become a crucial advocate of the feed-in tariffs proposed by Hans-Josef Fell, having originally favored a grant-based approach. A powerful orator and never a great compromiser, Scheer picked up several nicknames, including the "solar pope" or the "Stalin of renewables," depending on whether you supported his position or not.

Hermann Scheer lobbied for the foundation of the International Renewable Energy Agency and was president of Eurosolar, the European Association for Renewable Energy. His achievements have been recognized by many awards, including the Böer Medal[qv].

Latterly, he continued campaigning fervently for renewables right up to his death in 2010, after the publication of a book subtitled *100% Now: How the Complete Switch to Renewable Energies Can Be Achieved* had just been published [145], the last of a series [146,147] of influential volumes.

There is much more about Dr. Scheer's achievements in his own words in his chapter in the *Solar Power* book [43], and in the words of others in Bob Johnstone's book [82].

Dipesh Shah

Dipesh Shah brought the disciplines of professional business management to what became the world's leading PV company. Born in India, and raised in Uganda, he was educated at Warwick and London Universities and Harvard Business School. His career started in the oil industry, initially with Shell in 1974 and 2 years later at BP, where he held a variety of roles in London, New Zealand, and Scotland. Days after buying a house there at the end of 1990, he was asked to move south as CEO of BP's renewable businesses.

After a brief review, he concluded that its small PV business was the most promising renewable option, providing the best fit for the BP brand, particularly because it is suitable for use in urban environments and in the developed world. He reported accordingly saying BP Solar could grow into the world's leading integrated PV company while also becoming cash generative. Accepting his findings, the BP board said he could move on when those two goals had been met!

Shah promptly refocused the strategic direction of BP Solar. At the time, most of its business was for professional applications in the developing world, all off-grid and very much "feast or famine." By the time he moved on 6 years later, two-thirds of the market was grid-connected, mostly in the developed world. BP Solar was cash generative, and claimed the top spot for sales revenue – although it did not become the world #1 volume producer until the next year.

During this time, he was also active on the board of the European Photovoltaic Industry Association[qv], serving two stints as its president. Shah moved on within BP in 1997, and latterly headed up its acquisitions and divestment activities from 2000.

Shah in due course concluded he had a decision to make: Did he still want to be with BP when he retired? He decided not, and told his bosses he wanted to be away within the next 2 years. He was gone 27 months later. He then ran the UK Atomic Energy Authority for 3 years, perhaps the only person to have gone *from* solar *to* nuclear power.

Dipesh Shah has since served in many, mostly chairmanship roles, including some in the solar sector, with IT Power[qv] (2002–2004), NAPS[qv] (2003–2005), and Chinese PV manufacturer Jetion when it listed in London in 2007. He had been awarded an OBE earlier that year.

Ishaq Shahryar

Ishaq Shahryar was one of the most cheerful and courteous entrepreneurs one could wish to meet. Born in Afghanistan, he won a government scholarship to study in university in California, earning BSc in Physical Chemistry and Master's in International Relations – both disciplines he went on to apply.

Shahryar said he had "expected to go back to Afghanistan after my scholarship. Then in 1961 President Kennedy set the US on course to put a man on the Moon." He was captivated and joined on NASA's Jupiter Project, before moving to Spectrolab in the early 1970s, in the team that developed the process of screen-printing solar cells.

In 1976, Shahryar founded his own company, Solec International[qv], which was for many years among the world's top 10 solar producers. He always remained closely involved in the management and key account sales for Solec, where he was a formidable negotiator. His nephew Jawid worked with him from 1990, staying with the business after Solec was sold to Sanyo 4 years later.

Subsequently, in 2002, the former Afghan King asked Ishaq Shahryar to serve as the country's first Ambassador to the United States since 1978. His Excellency, Ambassador Shahryar served *pro bono* and invested his own money in the Washington, DC embassy.

After serving his term, Ishaq returned to California to start a new solar company, again with former Solec backer John Paul Jones Dejoria. Having been dubbed "the sun king" in a *New Scientist* article [148], he called his new company Sun King Solar. Following his death in 2009, Ishaq Shahryar was posthumously awarded a UN award as an "Inventor of Solar Technology."

Giovanni Simoni

Giovanni Simoni founded Italy's first photovoltaics business and Europe's PV industry association. Having graduated in mechanical and aerospace engineering, Simoni became assistant professor at Sapienza Università di Roma and was seconded as scientific attaché at the Italian embassy in London for 3 years from 1973.

He was already interested in solar energy, and in that year cofounded Solar Energy Developments, one of the UK's first solar businesses with Dominic Michaelis and Steven Szokolay. Its focus was on solar architecture, and one of the early projects was for a house in the hills outside Rome.

When he returned to Italy, Simoni took over responsibility for renewable energy at the Ministry of Industry, where he was involved with organizing the third international conference and exhibition on solar energy and rational use of energy in Genoa in 1980 and founding Italy's sustainable energy development agency, ENEA.

When the national oil company ENI set up its solar subsidiary Pragma[qv], Simoni was recruited as chief executive. During this time, he led the establishment of the European Photovoltaic Industry Association EPIA[qv], serving as its first president. He has also chaired the Italian solar association Assosolare.

After leaving in 1987, Simoni took on various national and international representative roles, including chairman of the solar expert group of the International Energy Agency and expert advisor to the Italian Banking Association and the European Bank for Development and Reconstruction. He held management positions for Gardini, Ferruzzi Montedison, and Nolitel.

Latterly, Simoni returned to the PV sector, establishing the advisory company that became Kenergia. He was president of WiseEnergy, its joint venture with Next Energy Capital. He subsequently sold Kenergia's O&M business to BayWa-re and has been appointed chairman of Moroni & Partners.

Scott Sklar

Scott Sklar is an astute political operator, who helped to steer the US solar industry through a difficult decade and more. In the mid-1970s, he was working "on the Hill" in Washington as an energy staffer, and advised the founders of the Solar Energy Industries Association[qv].

He then joined the National Centre for Appropriate Technology initially in Washington, and then as its Research Director in Montana. Scott moved on to the Solar Lobby – a coalition of green NGOs – as its political director.

In 1983, Sklar was invited to join SEIA as political director, and he took over as executive director 18 months later. This was a difficult period for the US renewables industry; Scott described his job [101] as "ensuring that Reagan didn't close everything down." Having worked in the senate, he knew how "politicians react to pain," and found himself not always popular with the administration or with those SEIA members who didn't want to rock the boat, reminding him of the hapless deer Hal in the Larson cartoon.

Scott left SEIA in 1999 after 15 years at the helm, although he did return as interim director for 6 months when his successor left. He then recruited Rhone Resch as a permanent replacement.

Sklar has subsequently been involved in other business and campaigning organizations, including the Business Council for Sustainable Energy and the Solar Foundation. He supported Michael Eckhart in founding ACORE, the American Council on Renewable Energy, to encourage the wider business community, such as financiers, to engage with renewables. Based in Washington, he now runs his renewable energy project business, The Stella Group, and teaches sustainable energy as a George Washington University professor.

Guy Smekens

Guy Smekens is something of a gentleman adventurer in a sector dominated by large corporates. After a spell in the nuclear industry, he moved into the consumer market with Proctor & Gamble. A chance visit to the laboratory of Roger van Overstraeten[qv] convinced him that PV was the field to work in, and he established ENE[qv] with two colleagues.

He was one of the founding directors of EPIA[qv] in 1985 and served as its treasurer for many years (Fig. 10.10). Smekens has also stimulated innovation in

Figure 10.10 Smekens presents accounts at EPIA AGM [68].

the sector through other investments, too. He was for a time a minority investor in HCT Shaping Systems[qv].

Guy Smekens has always enjoyed his pioneering work in the sector. His wife Simone, ever a strong supporter of the business, was nevertheless exasperated when, having successfully sold ENE, he agreed to buy it back from the German liquidator a few months later.

He has no plans to retire, saying "what better end for a farmer than to fall contented and dead into his ploughshare" [149].

Prof. Walter Spear

Walter Spear is credited with the invention of amorphous semiconductors, laying the foundation not only for solar cells but also displays and large-area electronics (Fig. 10.11).

Figure 10.11 Prof. Walter Spear [130].

He was brought up near Heidelberg, inheriting his scientific bent from his father, a pioneer in color photography, and his musical ability from his mother, a professional violinist. The Jewish Spear family escaped Nazi persecution in 1938, Walter arriving in the United Kingdom in his late teens with just a suitcase and a cello. After brief internment on the Isle of Man, he joined the Pioneer Corps, and later the Royal Artillery.

After demobilization, Spear obtained a physics degree and then a PhD in London, working on electro-optics. In 1953, he moved on to Leicester researching electron bombardment of dielectric layers. The most interesting results came from selenium films, where long lifetimes for free electrons and holes led to new avenues for research. Xerox Corporation was greatly interested in his work and appointed him as a consultant, as did EMI.

In the 1960s, he started working with research student Peter LeComber, who was to become his long-time colleague. LeComber later moved with him when he was appointed to the Harris Chair of Physics at the University of Dundee[qv] in 1968. They decided to focus on noncrystalline materials, about which they had been in correspondence with Sir Nevill Mott for some years, and started with amorphous silicon. In 1976, Spear and LeComber were first to publish research findings [131] on electronic and photovoltaic properties of amorphous silicon and germanium. They were acknowledged by Mott when accepting his Nobel Prize in 1977 for work on amorphous semiconductors.

Their pioneering work continued, attracting collaboration from industrial groups not just in photovoltaics but for wider electronics applications, until Spear retired in 1990, and LeComber's untimely death soon afterward.

Walter Spear was widely celebrated for his achievements. He was elected to the Royal Society of Edinburgh and later the Royal Society. He was awarded the Europhysics Prize, Max Born Medal, Makdougal-Brisbane Medal, Rank Prize in Optoelectronics, and the Rumford Medal. A great experimentalist and a gifted teacher, Prof. Spear died in 2008.

Prof. Richard (Dick) Swanson

Dick Swanson, an engaging astute and modest scientist, is another academic entrepreneur, and founded one of the few surviving early PV companies.

Having graduated from Ohio State University in 1969, Swanson moved to Stanford University for his doctorate in electrical engineering. He was working to reduce the voltage at which a CMOS (complementary metal–oxide–semiconductor) circuit could operate, from the 15 V norm at that time, using ion implantation.

The success of this approach surprised even Swanson, when he was demonstrating to a team from the US military, which funded the research. "I turned the voltage down to 1 volt and it kept running; half a volt – still good; then down to zero volts, and it was still going!" At first he couldn't understand it – and he could see the generals wondering if this was an elaborate con. Then he switched the lights off and the circuit stopped dead. The PV power from the ambient light on its p–n junction had been enough to keep the circuit going!

This was a relief to Swanson and his visitors, and a powerful introduction to photovoltaics. But it was the first oil crisis that really convinced Swanson that he wanted to work on solar power, rather than mainstream semiconductor research, which he felt was already getting "predictable." Having made the decision, finishing his thesis on ion implantation "became a slog" [70], and it never occurred to him to patent the process.

He joined the electrical engineering faculty at Stanford in 1976 as assistant professor, researching the semiconductor properties of silicon solar cells. He foresaw two routes to the substantial cost reductions then required: thin films, or smaller high-efficiency concentrator cells. As an expert in crystalline silicon, he chose the latter, devising efficient point contact techniques and other advances.

In 1985, Swanson incorporated what became SunPower[qv] with Dick Crane to commercialize their innovations. This led to a series of sabbaticals and leaves of absence, at one stage prompting the department chairman to ask: "Are you or aren't you a professor?" As SunPower's fund-raising efforts inched forward, Bob Lorenzini pointed out that financiers wouldn't want "a dilettante professor waltzing in once a week," and Swanson agreed he would join full time, when they secured the investment. That happened in 1991, and Swanson left Stanford to serve as SunPower's president and chief technology officer.

He has received many awards for his contribution to the sector, including the William Cherry Award[qv], the Becquerel Prize[qv], and the Karl Boer Medal[qv]. He subsequently reverted to part-time status at SunPower after its flotation, returning to Stanford. A fuller appreciation of his contributions to the sector can be read in his chapter in the *Solar Power* book [43]. Swanson retired from SunPower in 2012, and currently serves on various boards and advisory committees.

Takashi Tomita

Takashi Tomita was the chief executive responsible for Sharp Solar as it rose to become the leading PV company at the start of the new millennium (Fig. 10.12).

Tomita-san joined Sharp[qv] directly after graduating from Kyoto University in 1974. A specialist in semiconductor physics, solid-state electronics, optical processes, and microwave electronics, he contributed widely to Sharp's research and development activities across a range of fields: crystalline silicon cells and modules, amorphous and microcrystalline silicon thin film processes, III–V compound semiconductors, and even concentrators.

He progressed to become head of Sharp's Energy Conversion Laboratories in 1996 and was appointed general manager of the Photovoltaics Division the following year. In

Figure 10.12 Tomita and Scheer[qv] open Bavaria Solarpark [25].

that position, he was instrumental in leading the company to become the largest solar supplier in the world by 2000 – and for the next 6 years.

Takashi Tomita was also involved in several advisory roles beyond his executive positions at Sharp, including the International Electro-Technical

Commission (IEC), Renewable Energy Portfolio Standard of METI^qv, and NARA Advanced and Technology Institute.

Latterly, he was appointed professor at the University of Tokyo, established GENNAI, a think tank on harmonizing global environmental and energy issues, and established a CPV enterprise Smart Solar. He died in 2014 aged 63.

Bruno Topel

Having attempted to build a solar industry in South America single-handedly, entrepreneur Bruno Topel concluded he was the right person in the right place at the wrong time.

Topel's Polish family left Warsaw during the Second World War for Argentina, where he graduated in physics before moving to Brazil in 1964. He founded a company that made equipment for steel mills and became a multimillionaire, when he sold it to a Swiss multinational 15 years later.

While traveling the world with his wife and children, he happened to visit an energy exhibition in Australia, and was hooked by a new technology that turned the sun's energy into electricity. When he got back to Brazil, he promptly founded Heliodinâmica^qv, the first photovoltaic solar power company in the continent. A year and a half later, the factory started producing solar modules and 2 years later diversified up into cell processing. This was funded largely from Bruno's own resources, and those of his coinvestors. He personally handled most of the transactions with international suppliers in the United States and Europe.

While exasperated at the lack of coherent support from the Brazilian government, he was proud of the business he secured in the country. When the company delivered the first significant order to Embratel, he recalled "ministers announced how proud they were that such an important technology could be produced in Brazil by Brazilians. We at Heliodinâmica felt like national heroes!"

Subsequently, after investing some $10 million of his own money, about which he has no regrets, Topel decided enough was enough, and put Heliodinâmica "into hibernation."

He latterly advised mining machinery company Tecnometal on the establishment of a PV module plant and consults for other businesses in the Brazilian energy sector.

Frederick (Fred) Treble

Fred Treble's contribution to, and esteem within, the PV community stands in contrast to his country's modest involvement in the sector.

Treble started work as an apprentice on returning to the United Kingdom from overseas postings in an army family. He went to night school and completed a sandwich course to earn a degree in electrical engineering, later moving to the Patent Office. During the Second World War, he was at the Government Research Establishment and then the Ministry of Aircraft Production, joining the Royal Aircraft Establishment (RAE) after the war ended.

After working on missiles, Fred Treble moved to the PV section, at an exciting time with the rise of the space program in the 1950s. From 1960 to 1977, he was head of the solar group at RAE. It was active in several areas of PV solar energy conversion, particularly calibration and measurement techniques to improve performance and reliability. He was also involved with development and construction of lightweight arrays avoiding adhesives subject to radiation damage. He developed a method of calibrating reference cells and participated in international working groups to standardize test methodology.

In the late 1970s, Treble left RAE to establish his own consultancy business, focusing on the terrestrial application of photovoltaics. He led and participated in many of the international and national committees, which were working to establish crucial standards for testing and operation of PV devices and systems (see Section 5.6). Many of the resulting standards would have been much poorer and later, but for Treble's painstaking and courteous input.

He was a trusted advisor to the European Commission, a prolific author, and was involved in conference committees in Europe and further afield. He continued working until he was 90 and died 4 years later. At the 1994 World Renewable Energy Congress, he was recognized as a Pioneer in Renewable Energy and in 2000 he was awarded the Becquerel Prize[qv].

Prof. Roger van Overstraeten

Roger van Overstraeten was in all respects an aristocrat of the European solar sector. He gained his initial engineering degree at the Katholieke Universiteit Leuven[qv], and his doctorate at Stanford University[qv], returning to KUL after military service. He was appointed associate professor in electrical engineering the following year, and full professor 3 years later.

In 1975, he also took on a project overseeing the R&D program for the European Commission[qv], and was involved in organizing the first European PV Conference[qv] two years later. At about the same time, he was appointed to a similar role for the Belgian national energy program.

Prof. van Overstraeten was the prime mover in establishing IMEC (Interuniversity Micro-Electronica Centrum) at the end of 1983, and he served as its director until he died. IMEC was set up to enable universities in Flanders to collaborate in semiconductor research, sharing costs and facilitating transfer of technology to industry.

Throughout this period, his work covered many fields, relating mainly to silicon semiconductors. He was always keen to apply the resulting technologies to photovoltaic devices, often with coworker Robert Mertens. He studied the behavior of doped silicon and developed theories to enhance its use in transistors and solar cells, which have since been widely used in modeling electronic components. Van Overstraeten has many publications and was at various times visiting professor at the Universities of Florida, Stanford, and Pilani in India.

King Baudouin awarded him the title of Baron in 1989. The same year, he won the Becquerel Prize[qv], and he received many other awards up to and after his death in 1999. His commitment to knowledge sharing was commemorated by the creation in 2001 of the Roger van Overstraeten Foundation to help young people study science and technology.

Dr. Peter F. Varadi

Peter Varadi was a cofounder of the early leader of the terrestrial PV industry and is possibly the longest serving entrepreneur in the sector.

Varadi has written about his perilous escape from the Soviet suppression of the 1956 uprising in his native Hungary, shortly after he had gained his PhD in physico-chemistry from the University of Szeged. In 1968, he became the head of the chemistry laboratory for Communication Satellite Corporation in Maryland. Here he had his first exposure to PV, which Comsat used to power its satellites.

There he met Joseph Lindmayer[qv], his fellow countryman and opposite number as head of physics. They both had an interest in terrestrial PV and, having failed to persuade Comsat to support them, decided on New Year's Eve 1972 over several glasses of champagne that they should embrace the ideals of their adopted country, "become capitalists" and set up their own company. The first business plan was roughed out there and then, and the name – bowing to Lindmayer's preference for names ending in X – was chosen.

Solarex started on August 1 the following year, getting a big boost just weeks later, when the OPEC embargo sparked the first oil crisis. Peter and Joseph had already decided how they would divide responsibilities. Lindmayer ran science and technology, while Varadi volunteered [150]: "Since I had not yet won a Nobel Prize, I should get out of chemistry and go into some other field." He managed operations, including finances and business planning. "I didn't have any business training, but I had common sense, and am good with numbers."

When the company was sold to Amoco, Lindmayer went on to other things, but Varadi was retained as a consultant. He also established other interests, including the chromatography company Varex (Lindmayer's predilection for names ending in X had rubbed off!)

The European Commission asked Dr. Varadi in 1994 to head a study on financing renewable energy, with specific emphasis on photovoltaics. This highlighted quality assurance as one of the obstacles in financing renewable energy projects; so Varadi initiated the Photovoltaic Global Approval Program (PV-GAP, see Section 8.8) to test and approve PV products and systems.

He has latterly also advised the World Bank[qv], NREL[qv], and others. In 2004, in recognition of his lifelong service to the global PV sector, he was awarded the John Bonda Prize[qv]. Having written two books about the solar industry [51,151] (which tell his story better than I have here), Varadi is now turning his hand to fiction, to which he is well suited, he says, with his previous experience writing business plans.

Roberto Vigotti

Roberto Vigotti is a voluble enthusiast for sustainable energy, and has been a great promoter, especially for its use in the developing world.

After graduating in electrical engineering from the University of Pisa, he joined the Italian utility ENEL's management studies and research department in 1974 working on high-voltage transmission, moving in 1982 to coordinate research on new renewables, primarily solar and wind. ENEL started to undertake PV projects in the mid-1980s, initially mainly for off-grid applications, on islands, for example.

Vigotti was active beyond his industrial role, persuading the International Energy Agency to establish a PV Power Systems group, of which he was chairman for 5 years from 1990. He was also prominent in the International Solar Energy Society^{qv} and chairman of its Italian chapter.

Subsequently, Vigotti moved to business development for Enel Green Power in 2001 and then into international relations. He went on to other transnational activities on leaving Enel, with the IEA^{qv}, Paris-based Observatoire Méditerranéen de l'Energie and latterly the initiative for renewable energy in the Mediterranean, Res4Med.

Neville Williams

Because he's not a PV technologist, Neville Williams protests [35]: "I'm no solar pioneer"; yet he was among the first to introduce PV to off-grid communities (Fig. 10.13).

Originally a journalist, he was invited to join President Carter's newly formed US Department of Energy as a speech writer. He returned to his native Colorado after the change of administration, later returning to Washington as national media director for Greenpeace. This brought him into contact with many of the early PV companies, notably nearby Solarex.

Williams had traveled widely to developing countries and mentioned to Solarex CEO, John Corsi, that many people in those regions had no electricity, eliciting the simple response that "they also don't have any

Figure 10.13 Williams (on the right) in Nepal with Dak Bahadur [35].

money." Solarex nonetheless became the first contributor in 1990 to Williams' nonprofit Solar Electric Light Fund[qv] (SELF), giving him the impetus to raise further funding and introduce PV systems in 11 developing countries.

In 1997, the concept was extended with the establishment of a commercial enterprise, the Solar Electric Light Company (later SELCO Group). Neville acknowledges that he was not the first to introduce PV in the developing world, citing achievements of Richard Hansen in the Dominican Republic and Bernard McNelis[qv] in the nondomestic sector among others. He attributes the success of SELF and subsequently SELCO to the fact that, as a nonengineer, he focuses on application rather than on technology, and finds good partners to work with.

Subsequently, after retiring from SELCO in 2004, Neville Williams cofounded Standard Solar undertaking rooftop solar projects in the United States. More comprehensive and entertaining details of his contribution to the solar energy sector are available in his books [152,153].

Bob Willis

Bob Willis was the first professional manager brought in to lead an early terrestrial PV start-up. An army veteran and Boston College chemistry graduate, Willis was recruited from the aerospace industry by Exxon as president of Solar Power Corporation, when it moved into the commercial phase. Under his direction, SPC was the world's leading PV producer in the mid-1970s.

He left in 1978 to set up his own solar energy business, Solenergy, producing monocrystalline silicon solar panels in Woburn, Massachusetts. Solenergy never ranked among the leading world PV companies and was merged into Entropy Limited in 1984.

Willis subsequently became involved in town and state governments. He was also an expert and lecturer on the Second World War. Bob Willis died in 2015.

Philip Wolfe

One of the founders of the British solar power industry, Philip Wolfe obtained his engineering degree from Cambridge. After a couple of years at Raleigh Bicycles, he joined the automotive company Lucas Industries and was soon sent to research on the potential market for solar power in Canada. Knowing nothing about the subject, he visited Solar Power Corporation[qv] en route to the 3 month assignment. On returning to the United Kingdom, Lucas asked him to set up a subsidiary to promote solar opportunities internationally. In 1979, half of this company was sold to BP; to eventually become BP Solar[qv].

Frustrated that the parent companies were treating this more as PR than a serious business, Wolfe left in 1981 to set up his own company. With the support of three of Solarex's original investors – Roy Johns, Earle Kazis, and Joseph Lindmayer[qv] – he acquired Solapak[qv] from its then owners Solarex[qv]. He ran the company as chairman for the next 20 years, during the time it morphed into Intersolar Group[qv] and diversified into consumer product and thin film solar cell manufacturing.

Figure 10.14 Wolfe with Prime Minister Thatcher [27] opening the 1985 solar–wind hybrid project.

During this period, Philip led the Group in several innovative projects, including the UK's first grid-connected PV system and the first building-integrated solar–wind hybrid system in 1985 (Fig. 10.14). He was also an expert in system sizing techniques, with several publications [33].

Philip was a founder of the European Photovoltaic Industry Association and served as its third president; he was also a founder director of PV-UK.

Subsequently, following the proposed flotation and disposal of Intersolar Group in 2001, Philip served as Director-General of the UK's Renewable Energy Association, absorbing the British Photovoltaic Association PV-UK that he had cofounded years earlier, and leading the successful campaign for the adoption of feed-in tariffs in the United Kingdom. Thereafter he focused on utility-scale solar, writing a book [44], and founding wiki-solar.org in 2011; and on community energy, becoming the first elected chairman of Community Energy England.

Philip Wolfe was awarded MBE in 2016 for "services to renewable energy and the energy sector."

Prof. Masafumi Yamaguchi

Masafumi Yamaguchi has been working on PV research since the late 1960s, making significant contributions to the science of PV devices, mainly for space but also for terrestrial use.

He obtained his degree and doctorate from Hokkaido University. In 1968, he joined NTT Electrical's Communications Laboratory, and worked on radiation effects on silicon and III–V compound solar cells, initially as supervisor and later section head. He discovered the superior radiation resistance of indium phosphide, and was thus responsible for this material's first use in space on the Japanese satellite MUSES-A, which was launched in 1990.

In the late 1980s, Yamaguchi's group developed efficient multijunction devices with materials including gallium arsenide on silicon, indium phosphide, and tandem cells of aluminum–gallium arsenide with gallium arsenide.

Latterly, since leaving NTT in 1994, Yamaguchi has held several academic appointments and was appointed founding director of the Research Center for Smart Energy Technology at Toyota Technological Institute. He has continued to work on radiation resistance, discovered light-illumination-enhanced annealing in indium–gallium phosphide, and contributed to record-setting high-efficiency cells using compounds of gallium arsenide.

He led Japanese programs on Creative Clean Energy Generation using Solar Energy and Next Generation High Performance Photovoltaics and also the EU-Japan Joint R&D Project on concentrator photovoltaics.

He has subsequently supervised research projects that have set efficiency records for dye-sensitized and perovskite solar cells. Prof. Yamaguchi has received several PV awards, including the Becquerel Prize[qv] and William Cherry Award[qv], and served on many international committees.

John William (Bill) Yerkes

John William (Bill) Yerkes was one of the most inspirational and charismatic founders of the terrestrial solar power industry.

After an early career at Chrysler and Boeing, he started working on space cells in the late 1960s at the Textron subsidiary, Spectrolab, of which he became President and CEO. Soon after Hughes acquired Spectrolab, Yerkes was incensed to find himself "surplus to requirements" [154]. He had recognized the potential for terrestrial solar in the wake of the first oil crisis, so promptly founded Solar Technology International, funded by his severance pay, family, and friends.

The following year STI was acquired by the Atlantic Richfield company to become Arco Solar[qv]. Yerkes stayed on as Chief Executive, and started building the Arco Solar team in the Los Angeles suburb of Chatsworth.

His initial priority was to reduce the production costs of solar cells through more automated processes, such as screen-printing the contact grids, earning the title "Henry Ford of the solar industry" [155] He was a fountain of new ideas, most of which he wanted yesterday and whatever the cost. Researchers often found someone else was already on the project that Bill had just asked them to do.

Yerkes also concentrated on the development of an international distribution network, traveling the world to countless potential markets. He would often try to persuade those he met to "come and work with us." Yerkes' enthusiasm and energy were infectious, so several did, and that is how Peter McKenzie, among others, joined Arco Solar.

Arco Solar was a substantial diversion from Atlantic Richfield's core business. Yerkes had great respect for Ron Arnault, Arco's Director of Strategy and Planning, but according to reports [156], not all of the other managers drafted into its solar subsidiary appeared to have an affinity with semiconductor manufacturing. Yerkes served at times as Chief Executive, but had little sympathy for the bureaucracy of a growing business, and was at various times appointed President or Head of Development. Throughout most of this period, he managed his "skunkworks" – a small group working "outside the box" on innovative projects.

By 1985, Yerkes had become uncomfortable with his role in the company and left Arco Solar. It was an amicable arrangement and he frequently visited and kept in close contact with many former colleagues. He established Yerkes Electric Solar (YES) to work on cadmium telluride cells briefly, before returning to Boeing. Here he project managed the construction of the semiconductor fabrication plant in Bellevue near Seattle for III–V devices for radio frequency and optical applications and thin film solar cells.

Subsequently, Yerkes worked at Teledesic, a company founded by Bill Gates and Craig McCaw, to build 1000 Low Earth Orbit Satellites for Internet telecommunications. He later became Chief Technology Officer for Solaicx, which developed low-cost silicon solar cell processing, and was later acquired by SunEdison's parent company MEMC.

Yerkes, who died in January 2014, deserves much of the credit for many innovations in the sector, including automated processing and the first utility-scale solar power plant. But there can be no better testimony to his impact on those around him, and the solar industry in general, than the lapel badges worn at his memorial later that year: *Bill Yerkes changed my life.*

10.3 Not Forgetting

If I have seen further,
it is by standing on the shoulders of giants.

Isaac Newton [157]

Because of the selection criteria detailed at the start of this chapter, some notable pioneers are not profiled above. For a fuller list of people involved in the early PV sector, readers should peruse the full index at the end of the book.

Several giants of the photovoltaics sector are not included because our time frame starts only in 1973, or because their main involvement has been in the space PV sector. Details are available elsewhere of key contributions made by Martin Wolf [43], Gene Ralph,[1] Gerald Pearson [16], Joseph Loferski [6], Peter Iles [14], Calvin Fuller [16], William Cherry [6], Daryl Chapin [16], Henry Brandhorst [158], and Sheila Bailey [158].

There are people whose work made significant contributions to the PV sector even though this was not their primary career focus. They include William Shockley; Bob Chambers, and Ron Arnault at Atlantic Richfield; Eckehard Schmidt of AEG; Carl Weinberg at PG&E; and Mary Archer. Academic excellence depends on teams, so one should not overlook the contributions of Stuart Wenham at UNSW with Martin Green, or Robert Mertens at IMEC with Roger van Overstraeten among many others.

Similarly, in industry, significant contributions were made by many people who were not at the very top of their organizations, some of them are mentioned in Sections 11.2–11.5. Coming readily to mind in this category are Tim Bruton and Dick Evans at BP Solar; Dominique Sarti and Bernard Canal at Photowatt; Zim Putney at Solarex; and Paul Caruso, Ed Mahoney, and Bill Brusseau at SPC. Some people contributed in this way to the work of several solar organizations such as Dick Blieden at ERDA, Arco Solar, and USSC; Tom Dyer and Bob Kaufman at Arco then Photocomm; Peter McKenzie and Rob Muhn at Arco and AstroPower; Pierre Verlinden at KUL, SunPower, and latterly Trina; and John Wohlgemuth at Solarex, BP Solar, and subsequently NREL.

Then there are those who played important roles, but for only a minority of our time frame, namely, John Corsi and Harvey Forest at Solarex, John Wurmser at SPC, and Pierre Hamon at Chronar France.

In addition, there are those who started out in our time frame last millennium, but went on to make their major contribution subsequently – to be properly recognized, hopefully, should someone write the sequel to this book. Again, the list is long with obvious candidates including Frank Asbach who founded SolarWorld, which later acquired the residual assets of Arco Solar[qv]; Alf Bjørseth, who founded REC and Scatec; Bruce Cross at EETS and PV Systems; Chris Eberspacher, originally with Arco Solar and subsequently with Applied Materials and Hanwha; Jeremy Leggett, who did much to support the embryonic solar industry in his time at Greenpeace, and went on to found Solarcentury; Q-Cells cofounder Anton Milner; Georg Salvamoser, founder of Solar Fabrik; and Michael Schmela, cofounder and long-time editor of one of the sector's most read publications *Photon Magazine*.

10.3.1 Not Forgotten

Finally, I want to recognize again those who have not survived to see the full success of their efforts:

---

1 See Spectrolab in Section 11.2 and Refs [13,167].

Bill Yerkes[qv], Martin Wolf, Klaus Woerner, Bob Willis[qv], Fr. Bernard Verspieren, Roger van Overstraeten[qv], Fred Treble[qv], Takashi Tomita[qv], Harry Tabor, Walter Spear[qv], Ishaq Shahryar[qv], Hermann Scheer[qv], Georg Salvamoser, Paul Rappaport[qv], Iris Ovshinsky[qv], Stan Ovshinsky[qv], Hal McMaster, Arun Madan, Joe Loferski, Joseph Lindmayer[qv], Peter LeComber[qv], Karlheinz Krebs, Roy Johns, Matt Imamura, Bob Hill[qv], Walter Hesse, Brian Harper[qv], Yoshihiro Hamakawa[qv], Günther Cramer[qv], Dieter Bonnet[qv], John Bonda[qv], Werner Bloss[qv], Bud Annan, John Allison and others I have failed to recall.

My mission was to write this book before the list gets too much longer.

10.4 Where Did They Come From? Where Did They Go?

> "You can check out any time you want;
> but you can never leave"
>
> *Don Felder, Don Henley, and Glenn Frey [159]*

People have come to this sector from so many other walks of life, yet few move on after they've been captivated by the wonder of solar energy. Of those listed in this section, over 90% are still active in renewables – or they were until they retired or died. As Zoltan Kiss[qv] said: "Once you start working on photovoltaics, it is very addictive" or according to another old-timer [29] "like a virus."

10.4.1 Industrial Migration

Many people came into solar from the nuclear industry, fossil fuels, or utilities. Others came from the electronics and space sectors.

Perhaps the largest contingent were from nuclear, including Guy Smekens[qv], Wolfgang Palz[qv], Heinz Ossenbrink[qv], Bob Johnson[qv], Winfried Hoffmann[qv], Bob Hill[qv], Giuliano Grassi, and Joachim Benemann[qv]. The aerospace sector lost Bill Yerkes[qv], Bob Willis[qv], Peter Varadi[qv], Fred Treble[qv], Ishaq Shahryar[qv], Joseph Lindmayer[qv], and John Goldsmith[qv]. Petroleum industry migrants include Dipesh Shah[qv], Jeremy Leggett, Bob Kaufman, Brian Harper[qv], Jim Caldwell[qv], and Tapio Alvesalo[qv]; many of the French and Japanese solar pioneers came from electrical power and electronics.

Both Bill Yerkes, who returned to aerospace, and Dipesh Shah, who paradoxically moved on to the nuclear industry, kept their links with the sector.

10.4.2 Geographic Migration

When compiling the profiles earlier in this section, I was struck by the number of solar pioneers, who had been displaced from their homeland – often through necessity rather than choice.

The Jewish family of Walter Spear[qv] fled from Nazi Germany and that of Stan Ovshinsky[qv] from Lithuania. Bruno Topel[qv] left Poland during the war for South America. Joseph Lindmayer[qv] and Peter Varadi[qv] escaped Stalin's tanks rolling in to crush the Hungarian uprising; Zoltan Kiss[qv] emigrated a few years later. Ishaq Shahryar[qv] reluctantly left a war-torn Afghanistan behind, while Karl Böer[qv] quit Berlin when its Wall was built.

Others who made their name in PV far from the country of their birth are Engcotek's Ibrahim Samak, an Egyptian in Germany, and Dipesh Shah[qv], a Ugandan Indian in the United Kingdom.

I leave the reader to decide whether this was just a sign of the times in which the first solar generation lived, or whether disruptive technologies attract people who have themselves been displaced.

11

Profiles of Early PV Companies and Organizations

The following directory gives profiles of key organizations active in terrestrial photovoltaics during the early PV era. To make it easier to search, entries are grouped in the following categories:

1) PV manufacturing and research companies
2) Systems integrators, equipment suppliers, and other companies
3) Research centers and universities
4) International, national, and representative bodies

11.1 Inclusions and Omissions

I have tried to be objective and consistent in the criteria for including organizations in this chapter in the same way as for the list of people profiled in the previous chapter. Companies and research groups are included only if their innovations progressed to significant commercial exploitation during our time frame. I have excluded larger companies that dabbled in solar energy if their PV activities never emerged as a substantial business stream.

Because solar cells and modules were such a dominant component of PV systems in the early PV era, this chapter gives relatively comprehensive coverage to PV manufacturers in Section 11.2. Downstream companies in distribution and systems engineering were equally important to the success of the sector, but Section 11.3 is more selective in including only the most prominent. It also lists selected companies higher up the value chain, such as equipment suppliers.

Finally, the list deals primarily with organizations active internationally for a substantial proportion of our time frame. Regional players and later entrants, who became prominent only after the end of the last century, are generally excluded. A selection of those who missed out on inclusion because of these criteria is discussed in Section 11.6. Readers may choose to peruse the full index at the end of the book for a fuller reference of organizations and people active in and around the early terrestrial photovoltaics industry.

The Solar Generation: Childhood and Adolescence of Terrestrial Photovoltaics, First Edition. Philip R. Wolfe.
© 2018 by the Institute of Electrical and Electronic Engineers, Inc. Published 2018 by John Wiley & Sons, Inc.

11.2 PV Manufacturing and Research Companies

This section profiles companies whose primary involvement in the PV sector was in researching and producing solar cells and modules. Companies in the sector, whose main activities were not in the device research and manufacture are profiled in Section 11.3.

Remember: These lists contain just a selection of relevant organizations, and covers only the years 1973–1999. Later events are annotated "subsequently" or "latterly" or shown on timelines in italics. Superscripted "qv" relates to other people, organizations, events, or awards profiled in this Part II.

Refer to Section 11.6 for some obvious omissions.

AEG Solartechnik

What started as AEG Telefunken's solar department has been through so many changes of ownership and name, it is hard to know what to call it and where to put it! So, this profile deals with the origins of the business within AEG, and also the period after it had been sold to Daimler Benz and incorporated within their Deutsche Aerospace division (DASA). The later period after the merger with RWE's Nukem is covered under ASE below.

Summarized timeline

1967 Azur satellite uses AEG solar cells

1977 Terrestrial solar business starts

1982 Pellworm PV project (Fig. 11.1) – Europe's largest

1985 Daimler Benz buys AEG-Telefunken
      Solar goes to Deutsche Aerospace AG

1989 DASA acquires MBB merging solar business divisions

Figure 11.1 Pellworm plant [160].

1994   Solar business merged with Nukem's to form ASE with DASA
      holding 50%
      See separate profile under ASE

1996   DASA sells its 50% share in ASE

AEG-Telefunken's involvement in PV activity started when Germany's first research satellite Azur was equipped in 1967 with AEG solar cells.

The terrestrial business under the name AEG Solartechnik began in 1977 and received support for novel terrestrial solar cells and systems from the BMFT German ministry for research and development. The company also won backing from the European Commission^{qv}, including funding under their pilot projects^{qv} for an installation on Pellworm. The North Frisian island of Strand was devastated by a storm tide in 1634 breaking it into several smaller islands including Pellworm. The 300 kW solar system constructed by AEG in 1982–1983, under the project management of Dieter Mertig, was at the time Europe's largest PV array. Together with a wind generator it supplied a significant fraction of the power consumption of the island's 1200 inhabitants.

When Daimler Benz acquired AEG-Telefunken in 1985, it incorporated the solar business into its subsidiary Deutsche Aerospace AG (DASA), appointing AEG's Eckehard Schmidt as head of the Energy & Systems Engineering Division. It retained the terrestrial solar production plant at Wedel near Hamburg and space solar cell production in Heilbronn.

When DASA bought MBB (Messerschmitt-Bölkow-Blohm) in 1989, it merged its silicon solar activities with MBB's thin film (amorphous silicon and CIS) development, and these were later merged with the solar business of Nukem^{qv} to form ASE^{qv} (Angewandte Solarenergie), profiled below.

As a result of closure of solar activities at Wedel, former AEG Solartechnik employees founded various start-up companies in the area: Selected Electronic Technologies, Geosolar Energy and Environmental Systems, Gochermann Solar Technology, and Solarnova.

Arco Solar

Summarized timeline

1975   Bill Yerkes starts Solar Technology International (STI)

1977   Atlantic Richfield acquires STI

1979   Camarillo plant opened

1980   Becomes world's leading producer

1982   First megawatt-scale plant at Lugo

1983   Multimegawatt Carrisa Plain project

1990   Company acquired by Siemens

2002   Siemens Solar sold to Shell

2006   Shell Solar sold to SolarWorld

Arco Solar, the totemic solar company of the 1980s, was formed when in 1977 the Atlantic Richfield oil company acquired start-up company Solar Technology International that had been founded 2 years earlier by Bill Yerkes[qv].

Atlantic Richfield (Arco) had itself been formed through a merger in 1966, and its chairman Robert Anderson had a broad vision for developing the company. He was also a major owner of arid land and saw the potential for solar water pumping for his cattle. The solar venture was one of its most radical diversifications and Arco's Bob Chambers allowed Yerkes free reign to continue to develop the company, while assigning Peter Zambas to work alongside him (as "adult in charge" according to one employee). Zambas soon moved to oversee other Arco diversifications, and was replaced by Tom McLaughlin.

One of the first recruits at Arco Solar's premises in Chatsworth, California, was Charlie Gay[qv], who was appointed to build a pilot line and head up the research function. The initial mission was to develop solar cell processing better suited to high-volume production, applying techniques such as screen printing of the metallic contact grids, and later solar module lamination. Elliot Berman[qv] was brought in to support this work after he had left Solar Power Corporation[qv].

In the early years, the company's main focus was on building its market position, both by reducing production costs and by developing its distribution network. Arco Solar moved faster than most of its contemporaries to automate its production processes, with Terry Jester[qv] and others putting Yerkes' and Gay's ideas into practice. The Camarillo production plant was the world's biggest, and most advanced, when it opened in 1979 and the first to produce $1\,MW_P$ of modules the following year. It went on to achieve a cumulative total of $100\,MW_P$ by 1996.

Arco Solar's marketing policy focused on building key relationships in the world's leading PV markets under the guidance initially of Bob Kaufman brought in from Arco Chemical, then of Tom Dyer, Bob Nath, and Michel Rousseau. Arco Solar eventually had 96 distributors selling in 130 countries, and offices in Europe (Italy then London, under Maurizio La Noce and Ygal Giramberk), the Far East (Kevin Parnell), and Scandinavia (Jan Moe). Recognizing that systems sales were a key factor in developing its business, Arco Solar hired Peter McKenzie, who had been working on a solar-powered telecommunications project in Papua New Guinea and developed the applications engineering group with another former customer Raju Yenamandra. Recognizing the importance of transferring know-how in this emergent industry and the value of its employees in spreading information about its new products, Arco had a strong training program led by Mark Mrohs, and ably supported by Harlan Chapman and others.

Figure 11.2 Arco's 1 MW Lugo project [25].

These strategies enabled Arco Solar to build a dominant position in the world PV market, becoming the leading company between 1980 and 1989[1]. It was an exciting and inspiring company to work with or for in this era, still imbued with Bill Yerkes' vision and optimism. Its distributor conferences had an enthusiastic atmosphere akin to Apple's early new product launches two decades later. The company was also the first to build multimegawatt projects, with the 1 MW_P plant at Lugo (Fig. 11.2) in 1982 selling power to Southern California Edison followed by the 7MW_P plant at Carrisa Plain for Pacific Gas & Electric and further large-scale projects for Sacramento Municipal Utility District. Jerry Anderson, who had joined from Martin Marietta, played a key role in these developments.

Meanwhile, the research focus was shifting to new, potentially cheaper, solar cell options. Bob Chambers had retained Energy Conversion Devices[qv] to work on amorphous silicon, with a $20 million follow-on contract also involving batteries and other technology. In due course Arco also developed copper–indium diselenide (CIS) cells and hybrids of CIS with amorphous silicon.

The investments in production capacity, market building, and research ran to many millions of dollars, thanks to the enthusiasm and commitment of the head of Atlantic Richfield's strategic planning group, Ron Arnault. In the early 1980s, he was spending much more than the 2 days-per-week he was supposed to devote to Arco Solar, and had become its chief executive – Bill Yerkes having

---

1 Although some figures suggest it may have yielded top spot in 1986–1987 to Sanyo because of their thin film sales in the calculator and portable devices markets.

been appointed president. Recognizing that Atlantic Richfield would be increasingly reluctant to shoulder all this investment alone, Arco Solar entered into two separate partnership agreements, first with Showa Shell and then with Siemens, covering joint marketing and the further development of thin film solar cells.

Leon Codron and Peter Aschenbrenner[qv] were dispatched to support the Siemens joint venture, PV Electric, in Germany, and Tom Dyer, who spoke Japanese from his time in the US Air Force, to Showa. Showa focused on specialty products like calculator cells, Siemens on advanced concepts, such as solar sunroofs for cars, and Arco Solar on the tandem amorphous silicon/CIS cells.

The thin films made steady progress, used initially for consumer products and executive toys. In 1984, Charlie Gay burst in to the office of Jim Caldwell[qv], the then chief executive, to announce not another efficiency record – he'd done that several times already – but a stable amorphous silicon module that passed all the tests. Jim recalls [75]: "Now we actually had to step up and do something, but what? That was downright scary!"

Arco established a scale-up plant at Camarillo to manufacture its new Genesis module, no longer in the laboratory but under production conditions. Sadly, the pilot line proved incapable of consistently producing efficient cells at acceptable yield and throughput. Even though new efficiency records were set in the laboratory for amorphous silicon and CIS, Arco concluded in the late 1980s that it had no alternative but to advise its partners Siemens and Showa that it would be unable to deliver the technology that was expected.

This led in due course to Atlantic Richfield's decision to sell its solar business. Jim Caldwell resigned to put together a management buyout proposal and Charlie Gay was appointed president. Having considered the various proposals, the company was sold to Siemens[qv] in 1990. This led to a subsequent lawsuit in which Siemens claimed – unsuccessfully – to have been misled about the potential of Arco Solar based partly on a message found in their internal email system saying "as it appears that Arco's [thin film silicon] solar cell technology is a pipe dream, let Siemens have the pipe" [161].

In hindsight, it is perplexing that the world's most successful crystalline silicon producer should meet its demise in this way on the back of its thin film development program. Further details about the early exploits of Arco Solar can be found in Bob Johnstone's book [82].

Subsequently, Siemens sold its solar business to Shell[qv], which later divested most of what was left of Arco Solar, including the Camarillo plant to the listed German company, Solarworld, which in turn filed for insolvency in 2017.

Advanced Photovoltaic Systems (APS)

See Chronar Corporation and BP Solar

ASE (Angewandte Solarenergie – Applied Solar Energy)

ASE (Angewandte Solarenergie or Applied Solar Energy), was a joint venture formed in 1994 between Deutsche Aerospace AG (DASA, see above) and RWE subsidiary Nukem[qv] (see below), whose manager Winfried Hoffmann[qv] served as chief executive of ASE throughout its lifetime. The parent companies, which initially held 50–50 interests, merged their solar activities into ASE; in DASA's case, these included the former AEG[qv] solar business and that from MBB, together with Wolfram Einars, who joined ASE's management until DASA's later withdrawal.

ASE's first major action was the acquisition of Mobil Solar (formerly Mobil-Tyco), the leading producer of ribbon silicon solar cells using the EFG method, based in Massachusetts. This was renamed ASE America, and 3 years later Charlie Gay[qv] was appointed its president.

ASE's main European plant was at Alzenau near Frankfurt, with capacity about 20 MW incorporating the world's first fully automated inline production plant making EFG-ribbon wafers (Fig. 11.3), solar cells, and double glass modules (where the cell strings are laminated between front and back surfaces of glass).

ASE absorbed the Total-DASA joint venture Phototronics in the mid-1990s. In the late 1990s, DASA sold its holding in ASE to its joint venture partner leaving RWE with 100% ownership.

Subsequently, the company was renamed RWE Solar, until Schott bought a 50% stake in 2002 and it became RWE Schott Solar. Winfried Hoffmann left when Schott bought out RWE in 2005 and the company became Schott Solar. Schott eventually ended its PV and solar thermal activities in 2014.

Figure 11.3 EFG ribbon growth [25].

AstroPower Corporation

Formed in 1983 as the AstroPower Division of Astrosystems Inc., the business became an independent company in 1989, headed by Allen Barnett[qv].

Astropower started in contract research, funded primarily by agencies of the US government, and moved on to commence commercial solar cell manufacturing operations in 1988 using silicon wafers purchased from third parties. It soon began work on producing and sourcing lower cost silicon feedstock for its solar cell production.

In the 1990s, the company developed a process to recycle 6 and 8 in. round test wafers discarded by the semiconductor industry. These were cheap to buy and offered, when reprocessed, good quality for solar cell production. Alongside its monocrystalline process, the company was also developing its so-called silicon film, a process for depositing polycrystalline silicon in sheet form. These developments again benefited from significant assistance in the form of cost-sharing contracts from US government programs.

In the mid-1990s, AstroPower expanded rapidly hiring, amongst others, Peter Aschenbrenner[qv], Rob Muhn, Salama Nagib, and Peter McKenzie from Arco Solar[qv]. Howard Wenger also joined to lead the development of residential rooftop solar electric systems, distributed through partnerships with national home-builders, independent dealers, and the Home Depot national chain (Fig. 11.4).

The company went public in 1998 led by its Chief Executive Allen Barnett, President George Roland (who had joined from Siemens Solar[qv]), CFO Thomas Stiner and Peter Aschenbrenner. Overall, the company raised more than $100 million through private equity, corporate investors, venture capital, three public offerings, and bank debt.

Latterly, the company reached the world top five at a capacity of about 25 MW, and acquired the Spanish company Atersa in 2001.

Barely 2 years later, Barnett and Stiner resigned and the company filed for reorganization under Chapter 11 of the US bankruptcy code, at least in part to facilitate the sale of assets to GE Energy. This led to a rash of lawsuits [162]. GE Energy later sold the US assets to Motech, Elecnor bought the Spanish business, and AstroPower was liquidated.

Figure 11.4 Installing an AstroPower rooftop system [25] (health & safety standards were different in those days!).

BP Solar

Summarized timeline

1975   Lucas enters solar market

1979   Lucas Energy Systems formed

1981   BP acquires 50% of Lucas BP Solar

1983   Lucas sells out; now BP Solar

1996   Buys Advanced Photovoltaic Systems

1998   Amoco bought by BP
       Solarex absorbed into BP Solar

1999   World's leading PV producer

2011   BP closes all its PV business

From modest beginnings, BP Solar grew to become, at the end of the twentieth century, the world's largest photovoltaics company (Fig. 11.5).

The UK oil giant British Petroleum, which already had an involvement in solar thermal systems, took its first substantial step into the PV sector when it acquired a 50% share of Lucas energy systems in 1981. The company had been established within Lucas Industries, the UK-based automotive electronics multinational, by Philip Wolfe[qv], who served as first chief executive of the joint venture company, which was renamed Lucas BP Solar. BP bought the remainder of the company a couple of years later.

BP Solar was not at first a photovoltaic producer; concentrating instead on the downstream engineering and sales of systems for professional applications. It secured in 1980 the world's first two multimillion dollar PV contracts for rural telephony and telecommunications applications in Colombia and Algeria. It also undertook one of the EC pilot projects[qv], managed by Rod Scott, and installed on the coal yard of the defunct Marchwood power station beside Southampton Water.

Successive chief executives led BP Solar following Wolfe's departure, until Dipesh Shah[qv] was appointed in 1991. He refocused the company, which until then had operated as a start-up, to give it a more traditional commercial focus.

Figure 11.5 BP solar forecourt canopy [25].

The resulting emphasis on profitability made BP Solar perhaps the first major PV company to move away from routine annual operating losses.

The company started solar module assembly in the United Kingdom in the 1980s, but centered production in Spain in 1990. It forged a strategic partnership with Martin Green's[qv] group at the University of New South Wales[qv] and introduced its high efficiency Saturn range incorporating several of his innovations. New plants were opened in Australia and India, the latter in partnership with Tata.

Solarex[qv] and BP Solar were ranked as the world's third- and fifth-largest producers in 1998 when British Petroleum acquired Solarex's parent company Amoco, catapulting the newly combined BP Solarex (as they were called for several months, before reverting to BP Solar) to number one in the global PV industry. Harry Shimp was appointed chief executive of the merged business.

Under R&D Manager Tim Bruton, BP Solar collaborated widely on academic research and had interests in thin films even before the acquisition added Solarex's amorphous silicon. Another BP subsidiary Sohio had contracted Energy Conversion Devices[qv] to undertake thin film research back in the 1980s, and the BP Research Centre was working on cadmium telluride. In 1996, it acquired the Advanced Photovoltaic System's (APS) thin film plant in California, designed by the architectural practice of Gregory Kiss (son of Zoltan[qv]). APS had been founded to produce low-cost thin film cells, but had been restricted from amorphous silicon production because of Solarex' patent (see Section 5.7).

For a time, BP Solar continued the amorphous silicon work inherited from RCA through Solarex, and appointed Dave Carlson as chief scientist. However, this research effort was subsequently closed, as later was the cadmium telluride. Neither was sold, although both probably could have been.

Latterly, in the new millennium, BP reconsidered its expansionist strategy (it had temporarily even rebranded as "Beyond Petroleum") and BP Solar's star began to wane. The Australian and Spanish factories were closed in 2009, and US production the following year. BP Solar was finally wound up in 2011.

Chronar Corporation

Chronar Corporation was founded in 1976 by Zoltan Kiss[qv] in New Jersey to commercialize amorphous silicon thin film solar cells.

The company was aiming for costs at a fraction of the level at which crystalline modules were being marketed at the time – on the order of $10 a peak watt. It avoided the expense of continuous processing, adopting instead a batch system to deposit α-Si by chemical vapor deposition from silane gas. Individual cells were then separated as strips by scribing the deposited silicon layer, and a rear contact of aluminum was applied by flash deposition (Fig. 11.6). When produced in sufficient volume, this process proved itself capable of

Figure 11.6 Thin film plates in carousel for Chronar's flash deposition [27].

delivering low cost, but at relatively low efficiency – on the order of 5% – and inconsistent stability.

Initially, the company was privately financed, but capital raised through public offerings starting in 1981 enabled it to establish its own production plants in enterprise zones in New York, Alabama, and Trenton. It subsequently set up further plants through joint ventures overseas. Chronar France in Lens was the first of these in 1986. Chronar Corporation owned 49%, with the balance owned by various French and overseas corporates. Further overseas plants were established in Yugoslavia, in partnership with the electronics company Koncar; Wales in the UK[2]; Harbin, China; and Taiwan.

In the late 1980s, the Sheet Metal Workers Union invested in Chronar in return for it becoming a union shop. In due course the union brought in new management to improve financial performance, and Zoltan Kiss left. But a turnaround was not forthcoming and Chronar filed for bankruptcy in 1990.

For 3 years before its demise and several years thereafter, the combined output of the Chronar factories around the world provided the largest thin film production volume outside Japan. Some overseas factories were subsequently sold to new owners. Neste Advanced Power Systems[qv] bought Chronar France, reselling the thin film plant several years later to Dutch company Free Energy Europe. Intersolar Group[qv] bought the Welsh plant.

Back in the United States, the Union established a new wholly owned company called Advanced Photovoltaic Systems (APS), which lasted less than 2 years before its assets were acquired by BP Solar[qv].

DASA (Deutsche Aerospace AG)

Daimler Benz acquired AEG Telefunken in 1985 and incorporated the solar business into its Deutsche Aerospace division. Later, in 1994, the solar business was merged with RWE's Nukem to form ASE.

---

2 Unlike Chronar's other overseas ventures, this was wholly owned by Chronar, but it did receive significant grant support from the Welsh authorities.

Figure 11.7 Philip Lauwers at an ENE installation in Rwanda [163].

The period between and before these dates is covered under the profile for AEG Solartechnik above, the period after is covered under ASE above.

ENE (Energies Nouvelles et Environnement)

ENE is an independent PV company established in Brussels by Guy Smekens[qv], with former fellow students, Spigniew Szawlowsky, and René Grosjean in 1976 (Fig. 11.7).

The company focused on niche markets, mainly in Africa and Asia, and also supplied small cells for the hobbyist market through Radio Shack. Although never a major global player, ENE was always ready to innovate. It bought one of the first wire saws from Shaping Systems[qv], and was one of the first PV producers to make cells from 150 mm square wafers.

In 1990, the company entered the space cell business, applying its expertise in sawing thin wafers to gallium arsenide cells. ENE progressively withdrew from the terrestrial market; its last project was a solar minigrid for a 200 household island in the Philippines in 1998.

Latterly, the business was sold to a German company, which became insolvent soon afterward, and ENE was subsequently bought back by Guy Smekens.

ENEL

Italy's national electricity company, ENEL, was one of the first utilities to take an active involvement in photovoltaics, even though it considered the technology to be of little relevance for mainstream applications.

Renewable energy sources such as hydro and geothermal had been part of ENEL's business for a long time. In the early 1980s, despite the questionable success of its first solar project, a 1 MW concentrating solar power project in Sicily, a "new renewables" department was created to work on solar and wind power. When Roberto Vigotti[qv] was appointed, the department started work on photovoltaics, initially for off-grid applications, especially on islands, including

Figure 11.8 ENEL's Serre Persano plant after its expansion to 6.6 MW [45].

the Alicudi project under the European Commission's pilot program. Its first grid-connected rooftop project was in 1987.

ENEL was also one of the first companies to develop utility-scale PV power projects. In 1993, it installed a 3.3 MW project at Serre Persano near Naples, using solar modules from Ansaldo, Photowatt^{qv}, and Kyocera^{qv}. At the time, it was the world's second largest plant, behind Arco Solar's Carrisa Plain project, and it has latterly been extended to 6.6 MW (Fig. 11.8).

Subsequently, the group has created a subsidiary ENEL Green Power, which develops utility-scale renewable plants. This company has a substantial portfolio of projects, with a sizeable footprint in the rapidly emerging markets of South and Central America.

Energy Conversion Devices (ECD)

ECD was founded in 1960 in Michigan by Stanford and Iris Ovshinsky^{qv}, as a developer and producer of thin film solar cells, rechargeable batteries, and other renewable energy-related products. It was incorporated and taken public in 1964 to raise capital for its early work.

Initially, the company focused on researching thin film solar cells. It received high-profile contracts from Arco Solar^{qv} ($25 million) and BP's subsidiary Sohio ($80 million), amongst others. While these delivered development capital and resources for ECD, there were those in the recipient organizations who felt they had been oversold, receiving little value in terms of exploitable results.

ECD's later work on rechargeable batteries focused on nickel–metal hydride (NiMH) technology, originally developed for use in portable electronic devices, but becoming a serious candidate for use in electric vehicles. In 1994, General Motors acquired a controlling interest in Ovonics' battery development, patents, and manufacturing.

ECD developed its own range of thin film solar panels, produced by depositing the active amorphous silicon cells onto a stainless steel film in a roll-to-roll process, with the top surface protected by a plastic film. These were sold by subsidiary company United Solar Ovonic, and later under the Uni-Solar

Figure 11.9 ECD's Uni-Solar flexible solar panels [25].

brand by ECD's joint venture with Canon, United Solar Systems Corporation (USSC). In the 1990s, this venture was the world's largest producer of flexible solar panels, but rarely achieved profitability (Fig. 11.9).

ECD Ovonics was listed on NASDAQ from 1985, until subsequently in 2012 the holding company and its subsidiaries filed for bankruptcy.

ENI, Eurosolare

See Pragma, Italsolar, and Eurosolare (ENI).

France Photon

The French electrical conglomerate Leroy-Somer had been interested in PV since the early 1970s, partly because its subsidiary Pompes Guinard was supplying solar pumps in Africa and elsewhere (Fig. 11.10).

It became an investor in Solarex[qv] and established the jointly owned subsidiary France-Photon in 1978, managed by Alain Ricaud[qv]. It set up a plant in Angoulême employing some 30 people, producing mono- and polycrystalline silicon solar cells and modules, under Solarex' license. By 1980, total annual production volume was something over $100\,kW_P$.

The design and marketing of complete photovoltaic systems was undertaken by two other Leroy-Somer subsidiaries: The solar pump business was handled

Figure 11.10 Pompes Guinard solar pump in Mali in 1977 [25] using bought-in modules before France Photon started production.

by Pompes Guinard (with Father Verspieren's Mali Aqua Viva amongst their customers) while other market sectors were covered by Systemes Solaires. All these activities were supported by the French government and European Commission.

In 1984, a new systems company, Solar Force, was created by Leroy-Somer to commercialize the production of France Photon. However, this had barely started when the French government made clear that it intended to support only one major PV company in France, and France-Photon was merged into Photowatt[qv], with Leroy-Somer taking a minority holding.

Heliodinâmica

This entrepreneurial Brazilian company merits mention as one of the few PV companies outside the United States, Europe, and Japan. Indian producers CEL and BHEL were at times larger, but Heliodinâmica was more visible on the world stage.

The company was founded near Sao Paulo in 1980 by a group of private citizens led by Bruno Topel[qv], who together invested about $10 million. At first, it imported solar modules from Photowatt[qv] and Solar Power Corporation[qv], but in 1982 established its own processing capability, sourcing silicon from Wacker[qv] and equipment from Spire[qv]. By the end of the following year, with the help of technical staff from Sao Paulo Universities, the plant was running well.

Heliodinâmica supplied systems to local customers, such as the Brazilian public telecommunications company, Embratel (Fig. 11.11), and oil company, Petrobras. However, the company soon needed to look for international

Figure 11.11 Heliodinâmica solar power system for Embratel repeater [25].

markets to absorb its production volume. Having researched developing countries like Egypt, Kenya, and Argentina, it found it hard to compete at the module level with larger global competitors. Offering unpackaged solar cells, however, proved more fruitful, and the company achieved significant sales in India, and in more established markets like the United States, Germany, Italy, Spain, and the United Kingdom.

Plant capacity was expanded in 1985 and in 1986 Heliodinâmica became one of the world's top 15 producers. This was a substantial achievement, in the absence of any formal support program for PV in its home market; and facing the issues associated with operating in a developing economy – such as inflation at times over 2000%.

Heliodinâmica was a leader in the practical development of solar energy for use in developing countries. It produced a solar pumping system selected by the World Bank as early as 1981, and was always ready to discuss technology transfer to other emerging economies.

Subsequently, the company was "mothballed" in 2010.

Isofotón

Spain's leading indigenous early photovoltaics company, Isofotón, was founded in 1981 by Prof. Antonio Luque[qv], who served as chairman until 1989.

At its factory in Málaga it produced standard PV modules and also high-concentration systems. One of the company's specialisms was bifacial modules (Fig. 11.12). Isofotón also diversified into the production of solar thermal collectors from 1985.

In the late 1990s, Isofotón rose into the top 15 world producers, latterly, reaching the top 10 in the early 2000s. At its peak, it employed some 800 people with a presence in 60 countries.

Joint research remained an important part of the company's activities, working with universities and others in Spain and internationally, notably on silicon PV technology, concentrator modules, and tracking systems. Much of

Figure 11.12 Bifacial modules used for a sound barrier in Switzerland.

this research was supported by framework programs of the European Commission[qv]. The Bergé Group acquired Isofotón in 1997.

Subsequently, a minority stake was sold to Alba Corporation but later bought back by Bergé, which in 2010 sold the company to Affirma and Toptec. It closed in 2014, having been declared bankrupt.

Italsolar

See Pragma, Italsolar, and Eurosolare (ENI).

Kyocera

Kyocera Corporation – originally the Kyoto Ceramic Company – is a substantial Japanese enterprise whose main business is in ceramics, reprographics, and electronics. Led by Kazuo Inamori, it pursued primarily a commercial approach to the PV market, setting it apart from the majority of Japanese companies, which have tended to lead with government-supported programs.

Kyocera began research into solar power technology in 1975, at which time Inamori says "Supplying solar cells for widespread general use was a distant dream" [164]. It worked initially on silicon ribbon crystal solar cells using a sapphire substrate and was a participant in the Japan Solar Energy Company together with Sharp[qv], Matsushita, and US companies Mobil[qv] and Tyco, working to produce ribbon silicon technology.

Kyocera undertook its first overseas installation contract, powering a microwave communications station in Peru in 1979, the year it started commercial sales of PV modules. The following year, it established a new production plant in Yohkaichi. The company soon concluded that polycrystalline silicon would provide a surer route to low-cost production and started research on cast polysilicon in 1982, initiating full-scale mass production in 1986. Kyocera claimed the most efficient production solar cells, 15.1% for 100×100 mm cells, in 1987.

Its focus on commercial markets led to exports to many parts of the world, especially Asian markets such as Pakistan, China, and Mongolia (Fig. 11.13).

Figure 11.13 Kyocera PV system in China [25].

Having supplied panels to Switzerland for Project Megawatt initiated by Markus Real^{qv}, in the late 1980s, Kyocera installed its first grid-connected solar system in Japan in 1991, launching standardized residential solar power systems in 1993.

Kyocera became the world's No. 1 producer of solar cells in 1998. It has subsequently remained prominent within the PV sector, and also extended its activities into utility-scale projects.

Mobil Solar, Mobil-Tyco

See ASE (Angewandte Solarenergie – Applied Solar Energy).

Motorola

See Shell.

Nukem

This German technology company was established in 1960 by Degussa, Rio Tinto, Metallgesellschaft, and RWE, initially to work in the nuclear field, as its name (from Nuklear-Chemie und -Metallurgie) suggests (Fig. 11.14).

It later expanded its business into nonnuclear sectors, including photovoltaics. Nukem's first solar project involved the transfer to industrial production of Cu_2S/CdS thin film PV technology developed at Stuttgart University^{qv} by Werner Bloss^{qv}.

The company later decided to discontinue work on cadmium-based devices, and when in 1985 Winfried Hoffmann^{qv} (who had joined the company in 1979) was appointed to head the PV work, he focused on commercializing MIS inversion layer solar cells developed under Rudolf Hezel at the University of Erlangen and the Institut für Solarenergieforschung Hameln.

Nukem was also interested in amorphous silicon thin films and first pursued a prospective joint venture with MBB in the late 1980s. However, it was not until 1994 that the two businesses came together when ASE^{qv} was founded through

Figure 11.14 Containerized PV system in Tanzania [25].

the merger of the solar activities of Nukem and DASA^{qv}. This new venture, profiled above, was based at Nukem's headquarters in Alzenau.

Ovonics

See Energy Conversion Devices.

Panasonic

While not very active in the marketplace until toward the end of our time frame, Panasonic deserves brief mention because this period set the foundation of a business that, since the acquisition of Sanyo^{qv}, is an important player.

The company started research and development of amorphous solar cells in 1975, and began full-scale commercial production in 1980. It was also active in downstream applications, installing grid-connected residential systems in the 1990s under the third phase of Japan's Sunshine Project.

In 1997, Panasonic started mass production and sales of Heterojunction with Intrinsic Thin layer (HIT) modules, which had first been developed by Sanyo. Subsequently, Panasonic acquired Sanyo^{qv} in 2008–2010, and since then it has set records for solar cell efficiency.

Photowatt

Summarized timeline

1973	Philips RTC pilots PV cells and modules
1979	CGE establishes research activity
1980	Buys US-based Sensor Technology
1981	Photowatt formed by SAFT and ELF
1981	Philips merges its PV activities for 10% stake; production moves to Caen factory
1985	Photowatt absorbs France Photon^{qv}
1987	Chronar France acquires majority
1988	Management buys back control
1991	Shell^{qv} acquires 35% share Move to new premises near Lyon
1997	Photowatt bought by Canadian ATS
2012	*French assets acquired by EDF*

The French PV producer has had a long and colorful existence, partly because it has been seen by the government and others as *The* French PV producer.

Photowatt's origins stretch back even before 1979 when SAFT, the battery division of Compagnie Générale d'Electricité (CGE), established a small research team at its premises in the Paris suburb of Argenteuil, recruiting Dominique Sarti from Marseille University to lead the research. When the French government decided in 1980 to promote solar energy, CGE won the competition to select one nationally supported PV business, over France Photon[qv] and Merlin Gerin (now part of Schneider) amongst others. It needed to increase its capability rapidly and established a subsidiary in Arizona, which partnered with, and later acquired, California-based Sensor Technology.

Photowatt was incorporated in mid-1981, with the oil company Société Nationale Elf Aquitaine (SNEA or Elf) taking a 15% stake. CGE's Claude Remy[qv] was appointed as general manager and joined by colleague Robert de Franclieu[qv].

In the same year, Philips RTC, which had first piloted cell and module production back in 1973, merged its PV activities into Photowatt in exchange for 10% of the company. Emmanuel Fabre[qv] transferred from RTC to Photowatt as scientific director. Development and production, employing some 60 people under Doniminque Sarti, were relocated to RTC's premises at Caen. Commercial and systems activities – a further 40 people – were headquartered in Rueil-Malmaison in Paris.

The company grew steadily, majoring on opportunities in France and French overseas dependencies and former colonies. The majority of the business was for professional applications, notably telecommunications, with expertise in water pumping and other rural electrification projects (Fig. 11.15). In 1984–1985, Photowatt absorbed the activities of France Photon[qv], whose majority owner Leroy-Somer took a small stake in Photowatt. By 1986, the ownership was SAFT

Figure 11.15 Solar-powered pearl farm in French Polynesia [163].

43%, Elf 43%, Leroy-Somer 10%, and RTC 4%. However, SAFT's parent company CGE had now been privatized and were looking to divest unprofitable businesses.

This led to the ill-fated relationship with the French affiliate of Chronar Corporation[qv], Chronar France, which in July 1987 bought 56% of Photowatt from SAFT and Elf, who retained 15% each. Many employees soon became frustrated that the new majority owners had little affinity for Photowatt's core business but saw it as a vehicle to commercialize Chronar's relatively unproven thin film technology. Claude Remy left in October, soon followed by Robert de Franclieu. Emmanuel Fabre was appointed Chairman.

The foreign takeover of Photowatt also had critics in the French government, and irregularities in the transaction allowed them to encourage Remy and de Franclieu to lead a buyback. This difficult transaction was eventually completed the following year after all manner or political in-fighting, financial setbacks, and other adventures, more fully described by Remy elsewhere [143].

The next major expansion came when Shell[qv] subsidiary R&S wanted to increase its off take from Photowatt beyond their existing capacity, then less than 1 MW per annum, but still enough to make them Europe's largest PV producer. This needed new capital, which the company couldn't afford, so Shell invested in 1991 to obtain a 35% stake. A new production base was established in the municipality of Bourgoin-Jallieu near Lyon, which offered strong local support. Manufacturing moved from Caen to Bourgoin under the management of Charles Sgard. Most Photowatt modules now used polycrystalline wafers from a casting process developed by Dominique Sarti, and Photowatt sourced one of the first wire saws from Shaping Systems[qv].

Eventually, the management team concluded that to achieve profitability it needed to consolidate and reduce costs, but in doing so it would progressively lose its market position. The alternative of continuing expansion could only be achieved with new external investment, and in this they found a kindred spirit in Klaus Woerner, whose Canadian company ATS bought Photowatt in 1997. Sadly, Woerner died soon after, and Milfred Hammerbacher and Rusty Schmit assumed day-to-day management of Photowatt. Soon afterward ATS acquired the spheral cell technology (see Section 4.3) from Texas Instruments.

Subsequently, spheral technology failed to achieve commercialization and Photowatt's Lyon plant became part of the "PV Alliance" with EDF and others. In 2012, EDF Energies Nouvelles took control of all Photowatt's French activities.

Pragma, Italsolar, and Eurosolare (ENI)

The Italian oil company ENI, owner of the Agip petroleum business, has been involved in the photovoltaics sector since 1982 in several manifestations.

The company Pragma was inaugurated by Agip Nucleare in Nettuno south of Rome under the leadership of Giovanni Simoni[qv]. It was the first Italian PV

Figure 11.16 Giglio Island solar power system by Pragma [163].

producer, and also undertook research, development, and marketing of cells, modules, and systems, mainly at first in the Italian market.

The company developed two of the EC-supported pilot projects (see Section 13.1), one off the coast of Tuscany on Giglio Island (Fig. 11.16) and the other in Zambelli, Verona.

It later developed a casting process for lower cost solar-grade polycrystalline silicon through a subsidiary company Heliosil.

In 1987, control transferred to Agip corporate and the company changed its name to Italsolar. The solar business of Ansaldo was absorbed in 1992 to become Eurosolare, in which Ansaldo retained a minority shareholding until 1997. Following the privatization of ENI in the mid-1990s, it has taken over the ownership of Eurosolare from its subsidiary Agip. Throughout this period the company was one of Europe's largest PV producers, remaining just outside the world top 10.

Subsequently, the company enjoyed yet another brief name change to EniTechnologie and then in 2006 to EniPower, which continues to produce and commercialize solar modules and systems.

RCA

The Radio Corporation of America (RCA) deserves a brief mention because, although never a mainstream producer of terrestrial PV devices, it had such a pivotal role in early PV development and was a pioneer in amorphous silicon, one of the most promising terrestrial PV technologies.

RCA was one of the most influential electronics companies during the twentieth century with activities ranging from vinyl records to communications satellites. Paul Rappaport[qv] and Joe Loferski in the RCA Laboratories worked on the PV effect in cadmium in the mid-1950s and were encouraged by the US Signals Corps' William Cherry to start developing cells for space applications.

Twenty years later, RCA was the first company to make amorphous silicon solar cells under Dave Carlson[qv] and Chris Wronski. It successfully filed several patents [62] for its amorphous silicon process, and these later led to lawsuits, as mentioned in Section 5.7.

The RCA team set six consecutive records for the most efficient amorphous silicon cell before 1983, when it sold this research activity to Amoco, the parent company of Solarex[qv], who set a subsequent record. RCA had earlier set a record for monocrystalline silicon solar cell efficiency back in 1978.

The RCA Company was bought by General Electric in 1986 and was later dismantled.

Solar Energy Systems Inc. (SES)

See Shell.

Sanyo

The electronics giant Sanyo first got involved in PV when the Japanese government introduced the Sunshine Project in 1975.

It focused first, under the guidance of Yukinori Kuwano[qv], on amorphous silicon cells, sold under the brand name Amorton, initially in consumer products such as calculators. The range of applications was progressively extended to include flexible modules, building-integrated products such as semitransparent panels for windows, and even cells for a solar plane.

Sanyo was the first to deposit amorphous thin films on crystalline silicon cells to produce hybrid HIT (heterojunction with intrinsic thin layer) solar cells in 1990, claiming superior temperature performance. The company progressively evolved this technology, improving efficiency and reducing cell thickness. Development of the pure thin film products also continued, leading latterly to larger and tandem modules.

With Sumitomo Bank, Sanyo acquired the US producer Solec[qv] in 1994.

Late in the century, the company started designing the Solar Ark 3.4 MW photovoltaic generating system at its Gifu production plant, which was commissioned later in 2001 – at the time one of the largest in the world (Fig. 11.17).

Subsequently, following financial problems in the wake of the 2004 Chūetsu earthquake, which damaged its main semiconductor plant, Sanyo was acquired

Figure 11.17 (Half of) Sanyo's Solar Ark [25].

by Panasonic[qv]. Since this takeover was completed in 2010, the PV activities of the two businesses have been merged.

Sanyo and Panasonic have latterly achieved impressive progress with HIT cells, overtaking the pure monocrystalline record in 2014 with an efficiency of 25.6%.

Sharp

The Japanese electronics conglomerate Sharp can take credit as one of the world's first industrial solar cell producers, and has an extensive record in research, manufacturing, and applications engineering in mono- and poly-crystalline and thin film cells. It is one of the few companies in this list to have been involved before this book's "beginning" in 1973.

Sharp's founder, Tokuji Hayakawa, was decades ahead of his time in recognizing the potential for solar energy, reportedly saying back in the 1950s "If we could find a way of generating electricity from limitless solar heat and light, that would benefit humankind to an extent we can scarcely imagine" [165]. Hearing in 1959 that Sharp's research department had developed a working solar cell, he personally went to the small laboratory to congratulate and encourage his team.

Sharp launched a solar-powered transistor radio 2 years later. In 1963, it started volume production of solar cells and moved into professional applications with a solar-powered buoy in Yokohama Bay. Three years later it powered an entire lighthouse on Ogami Island with a 225 W array – then the largest in the world (Fig. 11.18).

In 1976, the company brought the first solar calculator to market, and in due course expanded its range of PV consumer products based initially on crystalline silicon.

The company innovated in both crystalline and thin film technologies under the direction of Takashi Tomita[qv]. In the early 1990s, it claimed the most efficient mono- and polycrystalline solar cells in volume production. Sharp started manufacturing roll-to-roll two-junction thin film devices in 1983, and has latterly set records for both dye-sensitized and amorphous silicon solar cells.

Figure 11.18 Sharp's first solar-powered lighthouse on Ogami Island [25].

The company diversified into residential grid-connected PV applications in the mid 1990s under the third phase of Japan's Sunshine Project. Sharp has innovated in balance-of-systems technology, too, with novel power conditioning and snow-clearance technology. It had also been involved in the space solar industry, with its cells first taken into space on the Ume satellite in 1976.

By the end of our time frame, Takashi Tomita had been appointed Group General Manager of Sharp's Solar Division and Sharp had again become the world's leading PV module supplier, a position it held until 2006.

Shell

Royal Dutch Shell had more forays into the early solar industry than any other company. As detailed in Section 5.9, Shell was the first major oil company to publish projections showing that it believed solar power would become a significant part of the energy mix in the future. Perhaps thanks in part to the pioneering work on peak oil by its researcher M King Hubbert.

Unlike most other oil companies active in the sector, Shell's initial activity was led by people from its chemical rather than the petroleum business, which perhaps influenced the focus on thin film solar cells. During the early PV era Shell made several sorties into the photovoltaic industry:

Showa Shell is separately described later.

SES Incorporated had been founded by Karl Böer[qv] in 1973 in Newark to work on a novel cadmium sulfide/copper sulfide solar cell, developed at Delaware University's Institute of Energy Conversion[qv]. It became a wholly owned Shell subsidiary the following year, under CEO Steve Dizio. The activity was well resourced with all the latest laboratory equipment. In its heyday, it had almost 200 people and a sales and marketing department, headed by Bob Johnson, although the company never made sales of more than a few kilowatts peak.

Despite achieving quite creditable conversion efficiencies, on the order of 10%, there were problems in reproducibly delivering the stability required for a commercial product. The company was closed soon after Shell and Motorola agreed to set up their joint venture:

Solavolt was a joint venture between Shell and Motorola, established in Arizona in the early 1980s. It developed and commercialized polycrystalline silicon modules, just reaching the world's top 20 producers before it was closed in 1987.

R&S Solar was a wholly owned subsidiary of Shell based in Helmond in the Netherlands. In 1984, it acquired the systems business of Holecsol, which had been founded 3 years earlier by Holec under a license from Solarex. Holecsol's CEO Jan van Rooijen also joined, and in 1986, R&S took over module production from its spin-off Zontec. It was renamed Shell Solar the next year.

Pilkington Solar coinvested with Shell to build a new solar cell factory in 1999 at the Gelsenkirchen site of its solar subsidiary Flachglas Solartechnik[qv].

Figure 11.19 Shell Pilkington solar factory in Gelsenkirchen [25].

This plant was amongst the largest in Europe. Pilkington had a 20% interest in the venture (Fig. 11.19).

Shell discontinued its solar module production in Holland after opening the new factory and completing the following US acquisition:

Arco Solar/Siemens Solar's legacy business and assets were purchased by Shell from Siemens in 2001–2002 (after our time frame for this book) and operated for a time as Shell Solar. Shell eventually sold the residual assets in 2006 to SolarWorld, which subsequently filed for insolvency in 2017.

Photowatt[qv], the leading French PV company, was minority owned – to the tune of 35% – by Shell between 1991 and 1997.

Other renewables promotional activities were undertaken from its London head office, headed between 1987 and 1996 by Roger Booth. Shell was active in promoting PV for appropriate applications, particularly in developing countries.

Altogether these initiatives must make Shell one of the largest investors in the early PV era, though it is not easy to detect a coherent strategic approach!

Showa Shell

Showa Shell Sekiyu was formed in Japan through the 1985 merger of two oil companies, Showa Oil and Shell Sekiyu, which had been closely linked for some decades.

The company entered the solar business soon afterward when it signed a joint venture agreement with Arco Solar[qv] to collaborate on thin film solar cell research and to promote Arco Solar's products in Japan. Arco's Tom Dyer, a Japanese speaker, was seconded to the joint venture.

The company embarked on an extended development program in collaboration with Arco (subsequently Siemens) Solar, academia, and the Japanese government. The research spectrum included crystalline silicon and amorphous silicon, but the primary focus from 1993 was on copper–indium diselenide (CIS). In common with other Japanese companies, Showa Shell spent years in the development phase before launching commercial production (Fig. 11.20).

Figure 11.20 The uniform appearance of Showa's CIS [25] suits architectural applications.

Subsequently, the company commenced commercial production in 2006 under the name Showa Shell Solar, rebranding as Solar Frontier 4 years later.

Siemens Solar

Siemens' involvement in PV was first initiated in the early 1980s within its nuclear energy subsidiary Interatom, led between 1982 and 1984 by Joachim Benemann[qv].

A dedicated solar subsidiary called Siemens Solar was founded in 1986 in Bergisch Gladbach, managed by Dietrich Stahl with Jürgen Duch as head of marketing. The company produced round monocrystalline cells and marketed PV generating systems. It was involved in several projects in the European Commission's demonstration program (see Section 13.3) including systems for the remote Greek islands of Antikythira (Fig. 11.21), Arki, and Europe's most southerly point, Gavdos. It also installed a PV-wind hybrid system for Sapientza lighthouse on Methoni Island.

Siemens collaborated on thin film cell development with Arco Solar[qv], establishing a subsidiary company PV Electric, where Arco's Leon Codron and Peter Aschenbrenner[qv] were seconded to work alongside Siemens' Hubert Aulich[qv]. This business concentrated on leisure applications, especially solar sunroofs for vehicles.

On the global scale, Siemens Solar was a relatively small PV producer, until in 1990 it acquired Arco Solar, whose president, Charlie Gay[qv], managed the merged company for the next 3 years. This acquisition instantly made Siemens

Figure 11.21 RF repeater on Antikythira solar powered by Interatom [25] using Siemens Solar modules.

the world's top PV company, a position it retained for much of the 1990s. In 1996, its (former Arco Solar) plant in Camarillo California became the first to reach a cumulative output of 100 MW peak.

Although there were prima facie synergies both with Siemens' power station business and its electronics activities, the merged solar venture did not prove a perfect fit for either. Responsibility was vested at first in the power division, but they lacked the expertise in volume manufacturing. Electronics experts were later brought in, with a positive impact on manufacturing and cost control, but less familiar with the idiosyncrasies of overseas off-grid markets.

In 2000, Siemens concluded it needed to concentrate its attention in its core business and started looking to divest the solar business, subsequently selling to Shell in 2001.

Latterly, Siemens has reestablished some involvement in the PV market, mainly aligned with its traditional business. It is involved in utility-scale solar, with the primary focus on inverter supply, electrical contracting, and operations and maintenance activities.

Solar Power Corporation (SPC)

Solar Power Corporation was the first company established in the United States specifically to develop solar cells for terrestrial applications, and the first to attract investment from the oil and gas industry.

Elliot Berman[qv] founded Solar Power Corporation in 1968 when he started to research lower cost solar cells. After renting laboratory space at Boston University, he secured funding from Exxon and moved to their New Jersey laboratory. A dozen research staff were recruited on Exxon's payroll.

The initial research was on organic solar cells with, at the time, about 1% efficiency. When Exxon questioned how this could lead to a viable venture, the team consulted other solar cell producers and concluded that the most credible business model was to find ways of reducing monocrystalline silicon cell costs to $10/W – from the $100+ typical space cells at the time.

It managed to achieve $15–20/W$_p$ using off-grade silicon wafers from the semiconductor industry and went into production. This required setting up a separate factory and sales office (in Braintree Massachusetts) in 1973. This had the effect of dividing the commercial business from research, despite Berman's reservations.

The first product was the P1002 module comprising five series-connected 2 in. round monocrystalline silicon solar cells (see Section 4.8). This module delivered about 1.4 W at nominally 2 V, and there was a lower voltage version, with the cells connected in parallel. The array in Fig. 11.22 has six subarrays each with eighteen P1002 modules.

SPC thus became arguably the first company after Sharp[qv] and Philips RTC to produce photovoltaic products specifically for terrestrial application. Amongst

Figure 11.22 Early SPC employees [25] include Elliot Berman & Ed Mahoney (center), Paul Caruso (right), and Art Rudin (left). Bob Willis was taking the picture.

the first recruits to the commercial business were Paul Caruso, Ed Mahoney, Bob Willis[qv] – Berman's former colleague at Itek – and long-haired Arthur Rudin.

A UK-based European subsidiary, Solar Power Limited, was established by Clive Capps, known to Berman from his pre-Itek days, and it later recruited Bernard McNelis[qv]. Bill Brusseau was appointed to head sales and marketing from the United States, and in 1978, Ed Mahoney opened a Houston office to address markets in the oil industry, including the Middle East.

The company progressed, along with the rest of the industry, to develop solar modules of nominally 14–16 V, suitable for charging 12 V lead–acid batteries. It established a capability in broader solar power systems and started building an international customer network.

There was nothing evangelical about Exxon's involvement in the sector, and they adopted a fairly hard-nosed commercial approach from the start. When Graham Lock, a senior director with one of SPC's biggest international customers Lucas, visited their headquarters on Avenue of the Americas in the late 1970s, he was told that Exxon Enterprises was "only interested in businesses worth over $1 billion a year, where we can maintain at least 30% of the world market."

It was soon becoming apparent that neither of these conditions were being met. When Exxon demanded that SPC aim for break-even, and refused Elliot Berman's request to make modest losses to fund continuing research, he left to return to academia. As time went by, cash-starved SPC was losing market share to Solarex[qv] and later Arco Solar[qv]. Bob Willis, who had been appointed CEO in the mid 1970s, left to establish Solenergy and was replaced by John Wurmser.

When the company was closed in 1983, Exxon made no attempt to sell the business as a going concern, even though there were potential purchasers, including an MBO involving Paul Caruso. Much of the manufacturing equipment was sold to Solarex. Further details about the early exploits of SPC can be found in Bob Johnstone's book [82].

Solar Technology International

See Arco Solar

Solarex Corporation

Solarex Corporation was one of the first four terrestrial photovoltaic companies founded in the United States in the mid 1970s. Joseph Lindmayer[qv] and Peter Varadi[qv], who were working in PV for space at the time, wanted to get involved in this embryonic market and proposed a new diversification for their employer Comsat. Unsurprisingly for a company called the Communications Satellite Corporation, the board were not interested in the terrestrial market and turned down their proposal; so Varadi and Lindmayer left to establish Solarex in 1973.

They were persuasive advocates of the potential for photovoltaics and, after rejections by venture capitalists, succeeded in raising seed capital from private angel investors, notably entrepreneur industrialist Albert Nerken, and in later rounds New York real estate developer Earle Kazis and Pittsburgh-based engineer Roy Johns.

Just weeks after Solarex started, OPEC sparked the first oil crisis – the company was in the right sector at the right time. They were brought back to earth with a bump early the following year, when Comsat brought a lawsuit for stealing its intellectual property. However, it soon became apparent Solarex was developing an entirely different process and the suit was dropped.

An early recruit in the research team, in Rockville Maryland, just outside Washington DC, was Lindmayer's former colleague and coauthor Chuck Wrigley. They started working on lower cost solar cells using, in common with their contemporaries, round crystalline wafers rejected by the semiconductor industry. Solarex was one of the first companies to increase the efficiency of its solar modules by cutting four edges off the round wafers to achieve a squarer shape, so reducing the lost area between cells. Ramon Dominguez joined in 1975 to oversee production operations.

Solarex was also one of the first companies to recognize that terrestrial solar cells do not demand the high levels of purity needed for integrated circuits, and could therefore use a lower "solar grade" of silicon. The company established a research program on semicrystalline silicon, which it called Semix, under Dr. Zimri (Zim) Putney. Its French partner Leroy-Somer agreed to provide the funding to put this into volume production [51]; and a European subsidiary Intersemix was established.

To help capitalize on the potential in global markets, Solarex set up a European office near Geneva in Switzerland, managed initially by Michel Joliot and later by Tom Rosenfield, and established joint ventures in several European countries. Two key collaborators were electrical equipment companies Moteurs Leroy-Somer in France and Holec in the Netherlands. Both invested in Solarex and established joint ventures – France Photon[qv] and Holecsol, respectively – in which Solarex held a minority share in exchange for technology transfer. A joint venture called Solaris in Italy was established with Montedison. Solarex's business in the United Kingdom, Solapak[qv], was a wholly owned subsidiary, because the local partner did not want joint ownership.

Figure 11.23 Solarex's Solar Breeder in Frederick, Maryland [25].

Solarex became the world's leading PV producer in the late 1970s. This high profile made it an attractive acquisition prospect, and a minority share was bought by Standard Oil of Indiana (known by their trading name of Amoco) in 1979, making the early Solarex investors the first solar generation's only PV millionaires.

Solarex recognized the value of automated production. It expanded its capacity in 1982 by building what it called the Solar Breeder, a new solar-powered solar module factory, in Frederick Maryland (see Fig. 11.23), financed by Amoco with building services designed in part by Solarex investor Roy Johns.

In addition to its mainstream PV systems business, Solarex sold offcuts of crystalline cells as microgenerators for consumer products. It wanted access to lower cost thin film technology, which was encroaching into this market niche, so Solarex agreed to acquire the amorphous silicon research from RCA[qv] in 1983. The unit was located in separate premises in Pennsylvania, funded initially directly from Amoco, and it progressively improved stabilized efficiencies for tandem cells to about 6%. Solarex started selling calculator cells through its Hong Kong office and actively defended its rights to the RCA patents.

By this time Solarex had more then 600 employees on four continents, but its growth was consuming a lot of cash. Amoco concluded that "it's difficult to go on funding them 100% when you only own 30%" [166]. Soon its increasing equity triggered a condition in the investment agreement that Amoco buy out the other shareholders, though this time the share price was not so attractive.

Amoco appointed John Corsi as chief executive, and Lindmayer left to set up Quantex and its sister companies in 1984. Varadi (whose own book [51] and his chapter in the *Solar Power* book [43] give great insight into the early Solarex years) also moved on to other things, while remaining an adviser to the company. Solarex lost the number one position to Arco Solar, as Corsi focused on improving its financial performance.

In the mid-1990s, Harvey Forest was appointed CEO, and Amoco teamed up with US energy company Enron to create Amoco/Enron Solar to build a thin film plant and commercialize this technology. Enron, in particular, was keen to go into volume production of thin films and the partnership started to establish a plant in southern Virginia.

Everything changed again when Amoco was acquired by British Petroleum in 1998, and Solarex was merged with BP Solar, initially under the name BP Solarex with Frederick as its head office. BP Chief Executive John Browne appointed Harry Shimp, who was at the time being recruited by Solarex, to lead the combined operation. BP continued to support the amorphous silicon work for a time, although it had its own research activity in the competing cadmium telluride technology.

Solarex latterly ceased to exist when BP Solar was closed, and sadly the Solar Breeder was demolished.

Solec International

Solec was established by Ishaq Shahryar^{qv} in 1976 as the fourth of the pioneering independent companies set up in the United States shortly after the first oil crisis (after Solar Power Corporation^{qv}, Solarex^{qv}, and Solar Technology International^{qv}).

Maybe because it was not acquired by a large multinational until much later than its contemporaries, it never became one of the world's leading PV producers, although it was often in or around the top ten. The company concentrated on volume production of modules using monocrystalline silicon solar cells from its factory in Los Angeles. It was also able to produce custom panels to meet the requirements of individual customers.

Solec did not undertake pure research, and was not involved in developing thin film materials, but focused instead on achieving competitive costs through efficient production processes and low overheads (Fig. 11.24).

Shahryar was able to finance expansion by raising funds from private investors, most notably the hair products entrepreneur John Paul Jones Dejoria.

The company was eventually sold to Sanyo^{qv} in 1994 but it did not maintain an independent identity, and Shahryar retained the rights to the Solec name.

Spectrolab

Spectrolab's PV activities have been mainly in the space sector but merits mention for its activities in, and influence on, terrestrial photovoltaics.

Figure 11.24 Car powered by Solec cells [25].

Founded in 1956 to work on optical devices, Spectrolab was one of the first to supply solar cells for space use through its Heliotek division, starting with Pioneer I in 1958 and making the solar array that Apollo 11 left behind on the moon. It continues to be an active participant in the space solar industry, setting record cell efficiencies.

Because of its work with NASA, Spectrolab received a contract from NASA Lewis to develop solar panels suitable for terrestrial application in the early 1970s. NASA wanted these, among other things, to provide power for communities around its tracking station in West Africa. Spectrolab was, therefore, responsible for some of the early work on solar cell encapsulation using both silicone resins and lamination.

Much of the early work on lower cost cell processing techniques, including screen-printing of the contact grids, was initiated at Spectrolab, before being taken on by Solar Technology International and Solec^{qv}. It participated in the JPL Block-buy programs designed to stimulate the early US terrestrial PV industry. Spectrolab was also prominent for its solar simulators used to test the performance of solar cells and modules.

The company was headed up by Gene Ralph and subsequently Bill Yerkes^{qv}, before it was sold by Textron to Hughes Aircraft Company in 1975. Several of the other pioneers of the terrestrial solar industry, including Ishaq Shahryar^{qv} and Charlie Gay^{qv}, also came out of Spectrolab, which has subsequently been acquired by Boeing. Its contributions to the PV sector are more fully described in Gene Ralph's acceptance speech for the Cherry Award [167].

SunPower Corporation

SunPower has had a long and varied existence, since its foundation in 1985 by two Stanford scientists, Dick Swanson^{qv} and Dick Crane as Eos Electric Power. The company's aim was to develop competitive high-efficiency concentrating solar systems.

For the first 2 or 3 years, it undertook only research projects, mainly for the Electric Power Research Institute. EPRI introduced experienced silicon wafer entrepreneur Bob Lorenzini, to the founders, and he joined the company as chairman, promptly changing the name to SunPower Corporation. After months of trying, the company raised funding in 1991 from two venture capital firms to establish cell fabrication facilities in a class 100 clean room in Sunnyvale. The first products were $1\,cm^2$ concentrator cells made using 3 in. round monocrystalline silicon; 52 cells per wafer. At this time, Dick Swanson left Stanford and joined the company as full-time president and chief technology officer.

Although the original plan had been to produce and sell complete systems, most of SunPower's early business was for cells alone. Because the technology had been developed for concentration, with sophistications like the interdigitated back contact, it was more efficient than most other terrestrial cells available at the time. This made it ideal when customers needed the maximum possible output from a limited area. The first significant order was from the Solar Systems company in Australia needing 200 cells and following up with an order for 1000.

In 1993, Honda was designing a car for the solar race in Australia and wanted 7000 high-efficiency cells. This was funded from their Formula One budget – money no object – so SunPower said it could deliver! Because of the high volume required, and because this was not a concentrator application, the company designed a new larger cell using half a 4 in. wafer.

The biggest problem, as Dick Swanson recalls, [70] was the large quantity required to a tight deadline, "We were academics and had no clue what to do." Luckily, he met former Stanford colleague TJ Rodgers, of Cypress Semiconductor, who in turn introduced a wafer fabrication expert, who agreed to spend the summer helping them fulfill the order. The factory geared up to three shifts and the cells were delivered. Honda won the race and invested in SunPower.

With little continuing demand for efficient but expensive solar cells, SunPower kept the factory operational by selling into other optoelectronic applications, particularly infrared detectors for barcode readers and data transfer. The company had an abortive digression into space cells; one satellite using trial cells blew up on take-off. Then SunPower's silicon cells got qualified for space use just as the space PV sector was moving over to gallium–arsenide technology.

Sporadic terrestrial high-efficiency applications continued with SunPower supplying cells for solar-powered aeroplanes including NASA's Pathfinder and Helios, and the Solar Impulse (see Fig. 11.25).

NASA consequently encouraged SunPower to investigate the potential for lower cost high efficiency cells: could they produce at $10/W rather than the $60 prevalent at the time? Working again with Cypress Semiconductor, SunPower was surprised to calculate that costs as low as $3–4/W could be possible, given high enough production volumes. This led to the subsequent transformation of SunPower from its origins as a concentrator research laboratory to

Figure 11.25 Solar Impulse, driven by SunPower cells [25] over San Francisco Bay.

the volume producer of flat plate solar modules it later became. Early in the new millennium, Cypress bought the company and established volume manufacturing in the Philippines.

Latterly, Tom Werner was appointed chief executive of the company and Peter Aschenbrenner[qv] joined. SunPower was floated on NASDAQ in 2005 and Cypress distributed its stock to its shareholders. The company acquired Tom Dinwoodie and Dan Shugar's Powerlight in 2007. Total[qv] has subsequently invested in the company and become its largest shareholder.

For a fuller account of the evolution of SunPower and its periodic struggles to meet the payroll, see Dick Swanson's chapter in the *Solar Power* book [43].

Total Energie

The French oil company Total has been involved in the PV sector since the 1970s through several ventures, some of which continue.

Photon Power of El Paso Texas was based on a small group, which had worked on cadmium sulfide solar cells since the 1960s. Total bought a majority interest in the mid-1970s and established a pilot production plant in partnership with US glass company Libbey-Owens-Ford. However, it proved unable to scale up at acceptable levels of yield, throughput, and stability.

Research activities then moved on to cadmium telluride. In 1992, the business was acquired by Coors (who renamed it Golden Photon), but it was later closed in 1996.

Total Energie Développement has led the company's commercial undertakings in the sector, since Roland Barthez, then a young engineer, persuaded the oil company to enter the PV market in 1983. It has been active supplying systems particularly in rural applications such as pumping and electrification, with a major focus on French overseas departments and territories. Barthez went on to lead Total's solar business to the end of our time frame and beyond.

Tenesol was established also in 1983 by Total as a producer of solar modules mainly at first for marine and then building-integrated applications (Fig. 11.26).

Figure 11.26 Tenesol panels in its factory façade [25].

Subsequently, a 50% share was acquired by EDF in 2005 but it was later bought back by Total in 2011. The following year, the company was merged into Total's majority-owned subsidiary Sunpower Corporation[qv].

Solems and Phototronics were 50–50 joint ventures formed in 1988 by Total and MBB/DASA[qv] to develop amorphous silicon thin films for photovoltaic and optoelectronic applications. The French entity Solems had 48 employees based in Palaiseau, headed by Alain Ricaud[qv]. Phototronics Solartechnik (PST), headed by Eckhard Roelen, was based in Putzbrunn, where the proposed production plant was to be established. The venture had not achieved full-scale production, before PST was absorbed into ASE[qv] in the 1990s.

Further ventures were established by Total after our time frame:

Photovoltech was founded in 2001 to commercialize IMEC's thin silicon wafer process. IMEC[qv] retains a small shareholding, with the majority owned equally by Total and GDF Suez (subsequently renamed Engie).

Konarka is a US-based organic solar cell company in which Total acquired a minority interest in 2008.

SunPower Corporation[qv] is profiled above. Total acquired 60% of the Nasdaq-listed company in 2011.

Latterly, renewable energies continue to play an ever more important role within Total Energie Développement, headed since 2007 by Arnaud Chaperon.

Uni-Solar

See Energy Conversion Devices (ECD)

11.3 Other Companies in the PV Sector

The following companies were prominent in the solar power market in our time frame, but not primarily as solar cell and module producers; although some in due course diversified into PV manufacturing as described.

Alpha Real

See Markus Real in Section 10.2.

Crystalox

Crystalox is one of the few companies to have successfully remained high up in the PV material supply chain, without vertical diversification into cell and module production.

The business was established in 1982 in Oxfordshire UK. It was among the first to develop multicrystalline silicon technology on an industrial scale, concentrating

Figure 11.27 Polycrystalline ingot by Crystalox [25].

initially on equipment for ingot production. After a time, the company changed strategy, moving into the processing and sales of ingots, and discontinuing sales of equipment. It was the first to produce 66 cm ingots (Fig. 11.27).

In 1985, Crystalox was taken over by Elkem, and subsequently bought back in 1994 by a management team including Iain Dorrity, who had joined in 1986.

In 1997 in Germany, Hubert Aulich[qv] and Friedrich-Wilhelm Schulze founded PV Silicon, and worked with an equipment manufacturer to develop a wire saw. Two years later, Crystalox and PV Silicon entered a strategic partnership that led to a merger in 2002, with Aulich joining the Crystalox board.

The company subsequently listed on the London Stock Exchange's Alternative Investment Market (AIM) in 2007 as PV Crystalox Solar.

Flachglas Solartechnik (Flagsol)

Flachglas is listed in this chapter because it started in PV as an applications business, becoming involved in cell processing only later through its partnership with Shell.

The German glass company Flachglas AG was acquired by the UK's Pilkington in 1980. Its expertise in mirrors and structural glass led it into the solar industry in the early 1980s, and Joachim Benemann[qv] was recruited to head the "Flagsol" business in 1984. The initial focus (literally!) was on parabolic trough solar concentrators, not on photovoltaics. The company supplied mirrors for the series of SEGS concentrated solar power projects in California from 1984.

When CSP started to stagnate and the PV market was accelerating, Flagsol decided to diversify into photovoltaics, and commissioned a report on the market by Strategies Unlimited[qv]. To maximize the benefits of its parent company's expertise and avoid the most intense competition, it concentrated on building-integration, especially solar façades (Fig. 11.28).

Figure 11.28 4 kW solar facade by Flachglas on office building in Berlin [69].

Flachglas's first major PV contract was a façade for the office of the Aachen-based utility Stawag, which acted as the proving ground for a double-glass module design pioneered by the company.

Architectural applications of this type demand custom sizes, shapes, and configurations rather than standard solar modules, as produced by most PV suppliers. The Aachen project was inaugurated, in pouring rain, in May 1991. Building-integrated applications proved very popular with progressive architects and led to many other projects – see Joachim Benemann's chapter in the *Solar Power* book [43].

This success prompted the decision to integrate vertically into solar cell processing and electronics. Having no PV process expertise of its own, it concluded an agreement with Shell[qv] to build a new plant next to the Flachglas factory in Gelsenkirchen (see Fig 11.19). The plant was designed with a nominal capacity of 25 MW, of which about 10 MW was in use when it opened in 1993. The module plant produced mainly double-glass panels, but also had the capability of making standard modules of the laminated construction more common in the industry at the time. The company latterly inaugurated what was claimed as Europe's first fully automated module manufacturing line

In 1995, Flagsol broadened its upstream activities further by acquiring a 50% stake in the inverter company SMA[qv].

Subsequently, in 2000, Flachglas AG decided to concentrate on its core market, and agreed a management buyout under the name Flabeg of the solar and automotive mirror businesses, which were then separated. The photovoltaic unit was sold to the Dutch Scheuten Glass Group, which has since left the PV sector. The SMA shareholding was sold back, and the concentrated solar power engineering department was sold to Solar Millennium. The automotive division as well as the CSP mirror manufacturing were sold to other overseas investors.

HCT Shaping Systems

This small company first introduced the wire saw, which proved to be important for mass production of crystalline silicon solar wafers and, therefore, for the early PV industry.

Figure 11.29 Early Shaping Systems wire saw [163].

Shaping Systems was established in Lausanne Switzerland by Dr. Charles Hauser in 1984, after he had left Solarex[qv] subsidiary Intersemix. The new company developed an innovative saw designed to cut multiple wafers from silicon ingots simultaneously (Fig. 11.29).

These machines used a moving wire covered in abrasive slurry to cut the ingot. At the heart of the device is a very long single steel wire, 100–200 μm thick, wound repeatedly around equally spaced grooves in guide rollers. This forms a curtain of parallel wires through which the ingot passes, thereby being sliced into multiple wafers.

Latterly, Applied Materials acquired Shaping Systems' business in 2007.

Intersolar Group

Intersolar Group, like Flagsol, is listed in this section because it diversified into solar device production only later in its history.

It has its origins in the solar systems engineering company Solapak, which was founded in Newcastle UK by Roger Mytton, and subsequently acquired by Solarex[qv] to become a 100% subsidiary – not Solarex's preferred minority holding because of Mytton's religious principles. In 1981, the business was bought by (this book's author) Philip Wolfe[qv] with two colleagues from Lucas BP Solar, Alan Dichler and Amir Haq, and with Solapak's manager Graeme Finch. These directors funded the acquisition alongside external private investors including Earle Kazis, Roy Johns, and Joseph Lindmayer[qv], who joined the board.

At that time Solapak marketed stand-alone photovoltaic power systems for professional applications such as telecommunications, transport, and oil installations. The company designed, sold, and sometimes installed PV systems using modules initially from Solarex[qv] and later from Arco Solar[qv]. Its main markets were in the Middle East and Africa, although it did undertake the UK's first building-integrated solar–wind hybrid project in Milton Keynes, inaugurated in 1985 by Prime Minister Margaret Thatcher – one of the European Commission's demonstration projects (Fig 10.14).

Figure 11.30 Thin film plates in Intersolar's clean room [27]. Spencer Jansen (right) & the author.

In the 1980s, the board decided to add a more proactive business stream and diversified into solar consumer products. This demanded a totally different commercial approach, but the company was successful "largely because we didn't know we couldn't do it" according to Philip Bouverat, who joined the management team to lead consumer product marketing. Several million solar-powered ventilators, lights, fountains, and other gadgets were marketed under the name Intersolar, and the company rebranded as Intersolar Group. Still only a small proportion of its business was in the United Kingdom, and in 1992 the company won a Queen's Award for Export.

Many Intersolar consumer products used thin film amorphous silicon solar cells; so when Chronar Corporation[qv] collapsed in the early 1990s, Intersolar Group acquired its UK factory, in South Wales (Fig. 11.30).

Under John McNeil and later Spencer Jansen, Intersolar Group redeveloped the process and drew up plans for a new larger production line to be funded through an IPO. By the end of the century, Intersolar Group was in the world's top ten thin film solar producers.

Subsequently, the proposed IPO had to be aborted, because of stock market instability following the 9/11 attacks. The assets were later sold and the company closed.

IT Power

This engineering consultancy was crucial to the deployment of renewable energy in developing countries, in particular because of its work in support of aid agencies and other finance providers in this sector.

IT Power was founded in 1980 by Bernard McNelis[qv], Peter Fraenkel, and Anthony Derrick (Fig. 11.31). The IT in the name has nothing to do with computing; it derives from Fritz Schumacher's Intermediate Technology Development Group that supported IT Power's establishment, as 50% owner for a couple of years.

Thanks to the founders' prior involvement in the World Bank's renewable energy pumping project, it was one of the few consultancies with any knowledge

Figure 11.31 IT Power founders test pumps in the Philippines [68] (from left) Peter Fraenkel, Bernard McNelis, and Tony Derrick.

of renewables in developing countries. By the mid-1980s, it had grown to some 30 people, undertaking projects for the European Commission, various UN agencies, World Bank, World Health Organisation, USAID, and others.

Although based in the United Kingdom, the company's business was almost entirely overseas, winning it the Queen's Award for Export in 1989. IT Power established at various times activities in India (led by Terry Hart), West and Central Africa (Jean-Michel Durand and Jean-Paul Louineau), South East Asia (Amir Haq), the United States, and others. In 1985, the company undertook its first project in China for the EU, which led to many further assignments in China and Mongolia.

It was also active in the development of selected renewable energy products. It worked for WHO on solar-powered vaccine fridges, assessing systems from the Centre for Alternative Technology and others, and led the design of a tidal turbine. The board recognized that product development activities needed substantial development capital, so Marine Current Turbines was in due course spun off separately. Peter Fraenkel went with it to work with Martin Wright.

A majority share in IT Power was sold to Indian investors after our time frame and it was subsequently merged into ITPEnergised. Bernard McNelis and Anthony Derrick have retired from the business and sold their remaining interest. For further background on IT Power, see McNelis's chapter in the *Solar Power* book [43].

Neste Advanced Power Systems (NAPS)

This business stream of Neste, the Finnish national oil company, is listed here, as primarily a systems company, although it too started PV device manufacturing late in our time frame. It was formally established in 1986, but dates back to a strategy group evaluating new energy technologies, formed by Neste chief executive Jaakko Ihamuotila soon after he was appointed in 1980.

NAPS was set up as a business venture working with Neste's Battery Division under the management of Tapio Alvesalo[qv]. Its remit included advanced storage technologies, photovoltaics, and hydrogen and electric vehicle concepts,

Figure 11.32 Holiday cabin in Finland with NAPS solar system [163].

but the main commercial products were batteries and solar systems. Initially, it bought in solar panels for its projects from United States, European, and later Japanese producers.

NAPS supplied systems globally, with good penetration particularly in Africa, Asia, and Scandinavia. It established offices in Sweden (headed by Leif Selhagen recruited from Ericsson), Norway (Magne Vegel, from battery company Noack), the United Kingdom (David Spiers from Arco Solar[qv]), France (Emmanuel Fabre[qv]), Kenya (Jim Fanning), and Singapore (Heikki Tikkanen).

NAPS' main technical focus was on systems engineering within both the professional and rural electrification sectors, including standard systems for "cold chain" refrigeration to rural health clinics, and for lighting and power in holiday cabins (Fig. 11.32).

NAPS identified thin film PV as a promising new technology and launched an R&D program in its sister company Microchemistry under Tuomo Suntola, who had won prizes for his work on atomic layer epitaxy. The thin film team obtained good results with cadmium telluride, but found stability elusive. The company decided to focus on amorphous silicon and bought, the former French subsidiary of Chronar Corporation[qv], as a commercial outlet for its future technology.

When Neste subsequently merged with the Finnish national utility IVO to form Fortum, it decided to seek partial divestment of NAPS. It was incorporated as a separate company in 2000, and nonexecutive director Dipesh Shah[qv] assisted the search for partners. Latterly, a full management buyout led by Timo Rosenlöf was agreed. The company NAPS Solar Systems continues to offer PV systems.

PV Energy Systems Inc.

After he left the Department of Energy, Paul Maycock[qv] established PV Energy Systems to advise businesses and policy makers on developments within the photovoltaics sector.

It published a regular newsletter "PV News," including an editorial from Paul under the byline "Boomer." The annual "World PV Markets" reports were based on confidential interviews of key players in the industry, and Paul also published a periodic "Photovoltaic Technology, Performance, Cost and Market Forecast." Most business plans for PV companies during the period would have incorporated input from PV Energy Systems and/or Strategies Unlimited[qv].

When Paul Maycock retired, the activity was sold to GTM Research.

R&S

See Shell in Section 11.2.

SMA Solar Technology

The inverter producer SMA Solar Technology (as it is now known) started life as the "Ingenieurbüro für System-, Mess- und Anlagentechnik" (engineering consultants for control, measuring, and equipment technology), spun out from Kassel University in 1981 by Werner Klein, Günther Cramer[qv], Peter Drews, and Reiner Wettlaufer.

The company was one of the first to develop and produce inverters specifically for PV systems and to incorporate peak power point tracking into inverters. It introduced the PV-WR in 1990, followed by the Sunny Boy in 1995 (Fig. 11.33). These early units were aimed particularly at domestic-scale systems of a few kilowatts, but the range later extended to the megawatt scale.

The company progressively increased the efficiency of its inverters from nominally 90% for the early models to the mid- and high 1990s latterly. It was also in the forefront of the introduction of string inverters, as an alternative to centralized units for large systems.

In 1995, Flachglas Solartechnik[qv] acquired 50% of the company, and Joachim Benemann[qv] joined the supervisory board. Subsequently, in 2003 SMA bought back this half-share and in 2008 it floated on the German stock market. Winfried Hoffmann[qv] sat on its supervisory board until recently.

Figure 11.33 SMA's Sunny Boy, the first mass-produced solar inverter [25].

Figure 11.34 SELF home systems in Nepal [35].

Solar Electric Light Fund (SELF and SELCO)

The not-for-profit Solar Electric Light Fund was founded by Neville Williams[qv] in 1990. It majors on the supply of solar home systems in developing countries. Typical early systems used 30 and 40 W solar modules and a battery to power fluorescent lights, a radio, and a small black and white television (Fig. 11.34).

Initially, SELF used funds granted by charitable foundations to buy these systems in bulk and sell them village by village in its target regions, usually in partnership with in-country nonprofit agencies. The financial model was a local revolving credit fund, whereby each customer household, often with support from neighbors in the village, made a down payment on the system and paid off the balance over several years. Repayments to the fund were then used to buy more systems, and so on, with a small portion used to establish a local dealership and train local installers and technicians.

This model was first launched in Sri Lanka and Nepal, and later adopted in over a dozen countries, including Zimbabwe, Vietnam, and China. In some countries, SELF evolved new project structures, such as the Indian joint venture, which formed a for-profit subsidiary with access to World Bank[qv] funds.

In 1996, SELF decided to spin off the for-profit Solar Electric Light Company (SELCO), cofounded in India by Neville Williams and Dr. Harish Hande. Williams relinquished his role at SELF to Robert Freling.

Subsequently, SELF enlarged its operations to rural schools and health clinics, water pumping and purification, and even small business applications. SELCO-India is still going and had installed a quarter of a million solar home systems by 2016. For a highly entertaining account of the trials and tribulations of the early years of SELF, readers should refer to Neville's book [153] and his chapter in the *Solar Power* book [43].

Solar Energy Centre

See Brian Harper in Section 10.2.

Solapak

See Intersolar Group.

Spire Corporation

Spire started as a research and technology business, but evolved in the late 1970s to focus on selling equipment and turnkey plants.

Roger Little[qv] founded Spire in 1969 to undertake consulting and research on energy beams and particle physics for the aerospace industry. By 1973, Spire had developed new manufacturing techniques for space cells, so was ready to participate as the terrestrial business ramped up following the first oil crisis.

Spire established solar module production and participated in the JPL Block-buy programs designed to stimulate the early US photovoltaics industry. Roger Little soon concluded that independent companies would find it hard to compete with the multinationals, which had started to dominate the industry, so he evolved the business plan to sell not product but know-how – production equipment, technology, and components. Spire floated its shares on the over-the-counter market in 1983, and Ghazi Darkazalli joined the following year from Exxon and was appointed vice president.

Spire started supplying equipment for measuring the performance of solar cells and modules at the end of manufacturing lines (Fig. 11.35). By the early 1980s, these Spire flash testers were used in many countries around the world. Progressively, the company diversified into a broader range of manufacturing equipment, the most widely sold of which were module laminators.

At the same time Spire stayed in the contract research business to enable it to keep up-to-date with evolving technology and resist obsolescence.

In due course, the company started to offer complete turnkey plants – typically capable of producing a few hundred kilowatts per annum. These enabled joint venture partners in overseas countries to establish themselves as local producers. Spire often contributed training in exchange for equity in the

Figure 11.35 Spire solar module test station [25].

partnership. When the plants became operational, Spire would supply the solar cells usually from leading US producers.

If volumes became sufficient, Spire could help local partners to integrate vertically up into cell manufacture, and it would then supply the wafers. It sold such solar module manufacturing plants to Brazil (see Heliodynâmica[qv]), Saudi Arabia, and India.

Subsequently, Roger Little retired at the end of 2013, having sold Spire's semiconductor business to Masimo Semiconductor the previous year. Dutch company Eternal Sun bought the solar simulator business in 2016 and, while the Spire still exists, it is substantially smaller than it was in its heyday.

Strategies Unlimited

Information provider Strategies Unlimited became the primary reference source on the development and status of the terrestrial photovoltaics industry [168]. Formed by John Day[qv] in 1979, its first contract from Paul Rappaport[qv] at RCA[qv] was to research the embryonic market for terrestrial photovoltaics.

As Strategies Unlimited gained clients and recognition, the team was expanded with Bob Johnson[qv] in 1982. Day and Johnson realized that this early-stage industry had few players with the resources to pay for expensive market research. Most of the permanent client base, therefore, would be multinational companies either active in, or interested in, the sector. Day describes Strategies Unlimited's role as "cheerleaders for the industry" [42] often brought in by PV divisional managers in large multinationals to persuade the board to establish (or stick with) a PV business stream.

Two of the Strategies' main regular publications were the "Solar Flare" monthly/bimonthly roundup of news and developments from the industry, and the annual shipment reports showing the output of each of the significant terrestrial photovoltaic manufacturers. The latter depended on regular contact with PV producers and distributors, giving the company close relationships with independent companies outside their core customer base.

By the late 1980s, Johnson was leading the PV work and Day expanded the company's activities more broadly into optoelectronics. Paula Mints joined in the mid-1990s to support the solar market research. Because accurate shipment reports were expensive to compile, the company needed to charge a realistic fee to a viable number of customers and to police unauthorized recirculation. Bob Johnson recalls having to discontinue the subscription of an unnamed Japanese government department because "we had only one customer in Japan, but every company there seemed to have access to our reports."

Strategies Unlimited also undertook custom market research assignments for clients on request. There must have been dozens of due diligence reports on significant investments during the early PV era, and very few would have been complete without a chapter from Strategies or PV Energy Systems[qv].

Figure 11.36 Wacker silicon feedstock and a finished ribbon solar cell [25].

On the retirement of John Day and Bob Johnson in 2001, the business was sold to Pennwell, which continues to serve the optoelectronics sector. After a break and a period at Pennwell, Paula Mints reestablished the PV business at Navigant Consulting, and subsequently set up her own independent business, SPV Market Research, which still publishes the Solar Flare and PV shipment reports.

Wacker Chemie

Wacker Chemie is a German-based chemistry and metallurgy company that has been involved in silicon material processing since 1959.

Its primary contribution to the terrestrial photovoltaics business has been the development of high-purity low-cost silicon feedstock through its subsidiary Wacker-Chemitronic (Fig. 11.36).

It was among a handful of companies successfully producing polycrystalline silicon (see Fig. 4.2) at industrial scale in the 1980s under Werner Freiesleben. It supplied this for both photovoltaic and other electronic applications. Latterly in 2007, it partnered with Schott to establish a new polycrystalline silicon plant in Germany.

Wacker continues to produce polysilicon in Europe and North America, with a series of capacity increases over the years. It also supplies encapsulation, potting, and bonding materials used in the PV sector.

WIP (Wirtschaft und Infrastruktur Planungs)

This Munich-based consultancy was established in 1968 by the Hammer family to advise on infrastructure in developing countries, as an adjunct to their architecture practice. It undertook projects in Africa, South and Central America in conjunction with the German development bank KfW.

Following the appointment of Dr. Peter Helm[qv] as its scientific director, WIP moved into photovoltaics and renewables in 1981, when retained by the European Commission[qv] to monitor its PV pilot projects (see Section 13.1).

The following year Dr. Matt Imamura joined from Martin Marietta in the United States and led much of the company's PV research work.

WIP took on the organization of the European PV Conferences[qv] from 1986 onward, a responsibility it retains to this day.

Customers and Early Adopters

It should not be forgotten how much the early PV industry was indebted to those early adopters and customers who took a chance with unproven technology, and in many cases helped the PV companies make their products more practical, competitive, and reliable. The companies listed here and in Section 3.2 are just a sample of those pioneers.

Early solar-powered navigational lights and lighthouses were pioneered first by the Japanese Coastguard, and then by Trinity House and the Canadian and US Coast Guards. Navigational aid suppliers such as Tideland Signal, Integrated Power (Fig. 11.37), Orga, and Automatic Power also became experts in solar system design and supply. They in turn were able to introduce solar navigational aids for offshore platforms in the Mexican Gulf, Indian offshore fields, and the Arabian Gulf. Abu Dhabi's offshore oil company ADMA-OPCO and India's ONGC were prominent users, and major offshore platform fabricators, such as Hyundai and Samsung Heavy Industries, were notable early adopters.

In telecommunications too, the solar industry owes much to end users prepared to adopt PV as a power source, including General Telephone Oman, Kenya Posts and Telecommunications, Petroleum Development Oman (part-owned by

Figure 11.37 Offshore solar installation by Integrated Power [25].

Shell^{qv}), and Telecom Australia. Here too, OEM system suppliers also took a lead in offering solar-powered equipment, led by AT&T, Bell Telephone Manufacturing Company, Cable & Wireless, EB Nera, GTE Lenkurt, Philips, Sirti and Telletra, with Racal in the defense sector.

Early adopters in the oil, gas, and pipeline arena included the national oil companies in Abu Dhabi (ADNOC), Iran (INOC), Kuwait (KOC), Qatar (QPC), and Saudi Arabia (Aramco), and the companies that supplied them, like Brown & Root, Mannesmann, and Snamprogetti.

11.4 Research Centers and Universities

Because so many research centers work on photovoltaic technology, I have had to be very selective. Only some of the most prominent national laboratories have been profiled here, however, the contributions of others, such as Sandia Laboratories, for example, are mentioned in Chapters 3 and 8.

Academia

In particular it would be impossible to do justice to all the universities who have contributed to PV research. They are, therefore, profiled below only if they have discovered new cell types, set efficiency records, or established early dedicated centers or institutes.

Many others have made equally significant contributions, including the institutions of academics listed in Section 10.2, such as Georgia Tech and Brown and Osaka Universities. The University of Kassel was responsible for the establishment of SMA^{qv} and set up in 1988 its Institut für Solare Energieversorgungstechnik (ISET), subsequently merged into IWES, the wind energy branch of the Fraunhofer Institute^{qv}.

Other academic research teams actively engaged include those of Gerry Wrixon and Sean McCarthy at University College Cork, John Allison at Sheffield, Bill Milne at Cambridge – and many, many more, not least those led by the many academic prize winners mentioned in Section 14.6.

University of Delaware – Institute of Energy Conversion

The Institute of Energy Conversion (IEC) was created back in 1972, before the first oil crisis, with funding from the National Science Foundation and electric power utilities. It was established to work on coupling thin film photovoltaic cells with thermal collectors to supply the energy needs of individual homes, as proposed by its first director, Prof. Karl Böer^{qv}. In 1973, the first house to incorporate solar energy for both heat and electricity, called Solar One, was built at the University of Delaware (Fig. 11.38).

Figure 11.38 Delaware's Solar One demonstration house [25].

IEC soon started research on thin film Cu_2S/CdS (copper sulfide and cadmium sulfide) solar cells and progressively increased efficiency from about 3 to over 9% by 1978.

In 1975, Karl Böer shifted his focus to SES[qv] and Dr. Allen Barnett was appointed as IEC's director a year later, serving until 1979, when he took up another position at the University and started working on AstroPower[qv]. The breadth of research activities was expanded to include other II–VI compound solar cells and amorphous silicon (α-Si). A program to develop commercial-scale thin film manufacturing processes was initiated, and later funded by Chevron.

In 1980, IEC developed the first thin film solar cell to exceed 10% efficiency, using a copper sulfide/cadmium zinc sulfide (Cu2S/CdZnS) structure. However, issues associated with the stability and encapsulation of this device led IEC to redirect efforts to amorphous silicon and copper–indium diselenide (CuInSe2 or CIS).

IEC moved to a new purpose-built laboratory in 1982. There it added cadmium telluride (CdTe) to the thin films it worked on, becoming the first laboratory to fabricate cells at over 10% efficiency utilizing four different absorbing semiconductors: α-Si, CdTe, CuInSe2, and Cu2S. In the late 1990s, it also started work on polycrystalline silicon cells.

Throughout its history IEC has worked with many companies, and several of its alumni have contributed to the development of the PV sector. The University of Delaware awards a Karl Böer Solar Energy Medal[qv] every 2 years (latterly annually) to recognize significant contributions to the sector.

Dundee University

Dundee University's work in PV-related areas began in 1968, when Professor Walter Spear[qv] was appointed to the Harris Chair of Physics, and his coworker Peter LeComber[qv] joined him. They decided that a main research aim of the new laboratory should be to experiment on the noncrystalline state, about which they had been in correspondence with Nevill Mott for some years. They

Figure 11.39 Early thin film transistor display using α-Si from Dundee [169].

started with amorphous silicon (α-Si) as a suitable model material, and soon found that deposition of α-Si films from silane in a radio frequency glow discharge offered promising semiconductor properties. Their results were later acknowledged by Mott in his 1977 Nobel Lecture.

Perhaps the most important breakthrough was achieved in 1975 when the Dundee group demonstrated that, contrary to the prevailing opinion, plasma-deposited α-Si, and amorphous germanium (α-Ge), could be doped to give it semiconductor properties directly from the gas phase during deposition. In 1976, Spear and LeComber published results on the first amorphous p–n junction electronic device shown to demonstrate photovoltaic properties [131].

The Dundee group continued research to enhance understanding of α-Si and α-Ge, and particularly the effects of hydrogen content and various doping approaches to the semiconductor properties of the material.

During the 1980s the study extended to the electronic properties of nano-crystalline silicon, which Japanese researchers had shown could also be produced by plasma deposition, as a function of crystallite size. They did further work on plasma-deposited device structures, concluding that there was no fundamental limit to the size of the deposited films; a factor that would be important in large area applications (Fig. 11.39).

The amorphous materials team produced other researchers, such as Arun Madan, who went on to other achievements in PV and semiconductor science.

The group duly found itself in demand from industrial research groups for collaboration on new device ideas. It worked extensively on nonphotovoltaic applications of amorphous materials for devices such as field effect transistors, and this in due course led to large area liquid-crystal color displays, and thin film memory elements.

École Polytechnique Fédérale de Lausanne (EPFL)

This Swiss university has two dedicated PV groups.

Probably EPFL's most notable innovation came from Laboratory of Photonics and Interfaces (LPI) headed by Prof. Michael Grätzel[qv], which works on

Figure 11.40 Lightly tinted dye-sensitized solar cells on EPFL's campus in the Swiss Tech building's facade.

photosystems that produce electric power or fuels from sunlight. It was here that dye-sensitized solar cells (see Section 4.6) were first shown to have potential for commercialization. These use surface-bound dyes on mesoscopic oxides as an electron capturing substrate, permitting design of relatively efficient photovoltaic thin film organic solar cells (Fig. 11.40).

A second related theme of research at LPI is the generation of solar fuels with photoelectrochemical systems that employ mesoscopic semiconducting oxides as light absorbers.

EPFL also hosts the Photovoltaics-Laboratory (PV-Lab) within the Institut de Micro-Technique (IMT). PV-Lab was founded in 1984 by Prof. Arvind Shah and was originally part of the University of Neuchâtel, but subsequently merged into EPFL. It has pioneered several new processes for the preparation of thin films, such as very high frequency plasma deposition of microcrystalline silicon and processing of heterojunction cells.

Fraunhofer Institute for Solar Energy Systems (F-ISE)

Founded within the Fraunhofer Society in 1981 by Adolf Goetzberger[qv], F-ISE in Freiburg grew to become the leading solar energy research center in Europe.

The Fraunhofer Society is a group of several dozen research institutes across the country, minority funded by the German federal and Länder governments. It agreed, with some reluctance, to establish an Institute for Solar Energy Systems (F-ISE) in 1981. This was directed by Adolf Goetzberger[qv], who transferred from another Fraunhofer Institute with some 18 staff.

The remit for F-ISE was intentionally broad, covering both devices and systems development. Jürgen Schmid joined from the European Commission[qv] in 1981 to head the system engineering department. Much of its initial work was on solar architecture and the institute went on to build Germany's first 100% solar house (with no connection to the electricity grid) in 1992 at Freiburg (Fig. 11.41).

The Fraunhofer model needs majority funding from industry so, as Goetzberger recalls [170], research on photovoltaics "in the early days was difficult, because there

Figure 11.41 F-ISE solar house in Freiburg [25].

was no industry." However, some projects were developed with Wacker[qv] and AEG[qv], and progressively the PV activities grew to become a significant part of the center's work.

Again, a whole system approach was taken and F-ISE worked on storage and electronics alongside PV devices, such as fluorescent collectors. The systems department developed arguably the first fully electronic inverter. Because of its systems expertise, the institute was contracted by the German government to monitor its 1000 Roof Programme (see Section 8.4).

By the time Professor Goetzberger handed over the directorship to Joachim Luther in 1993, F-ISE had grown to some 200 people, making it one of world's largest solar research centers. At about the same time, Jürgen Schmid left to join the University of Kassel, subsequently ending up back at Fraunhofer when Kassel's ISET merged with the Institute for Wind Energy and Energy System Technology in 2009.

Latterly, F-ISE has continued to thrive and weather the fluctuations in the European solar industry, and is now one of the largest in the Fraunhofer Society. Since 2006, its director has been Eicke Weber.

IMEC & Katholieke Universiteit Leuven

The Belgian University KUL (Katholieke Universiteit Leuven) started its work on semiconductors at the time Roger van Overstraeten[qv] was appointed professor in the late 1960s, and included photovoltaics among its research specialties soon afterward.

Van Overstraeten was instrumental in the creation of the Interuniversity Micro-Electronica Centrum (IMEC) at the end of 1983. IMEC was established with an initial grant of €60 million from the Flanders government to enable universities in the region to collaborate in semiconductor research, sharing costs, and facilitating the transfer of technology to industry. At the same time, a semiconductor factory (MIE-TEC) was also founded.

Within a few months, 70 staff had been appointed working from premises at KUL. An associated laboratory was established at Ghent University, working on chemical vapor deposition of III–IV materials. In 1986, a new "clean room" was opened at IMEC, inaugurated by the Belgian King. This facility was used for the fabrication of 125 mm wafers. The company Soltech was spun out 3 years later to commercialize PV systems based on IMEC's silk-screen process.

In 1994, IMEC in partnership with ESAT and Philips cofounded a development program for digital signal processing technology, and it has latterly continued to collaborate with others as one of the country's leading research establishments for semiconductor and PV technology. Several KUL alumni have moved into prominent roles in the PV sector, including Pierre Verlinden who worked at SunPower[qv] and Trina.

Jet Propulsion Laboratory

Space applications have of course been the primary focus within the photovoltaics sector of NASA [158], but its Jet Propulsion Laboratory (JPL) merits brief mention here for its role in early terrestrial PV.

The Jet Propulsion Laboratory acted as the independent test center for PV modules because it was, in the early years, one of the few national laboratories with experience in the technology, and it participated in the international "round robins" on PV testing described in Section 5.6.

JPL, which is managed on behalf of NASA by the California Institute of Technology (CalTech), also coordinated the Block-buys and participated more broadly in the US Department of Energy's solar development program (see Section 8.3).

Joint Research Centre – Ispra

The Institute for Environment and Sustainability (IES) is one of seven scientific institutes of the European Commission's Joint Research Centre (JRC) located at Ispra in Italy. JRC brings together multidisciplinary teams to provide scientific and technical support to the development of European policy.

Under the direction of Gerd Blaesser, the renewable energy work was led in our time frame initially by Karlheinz Krebs and later by Heinz Ossenbrink[qv], supported by Ewan Dunlop. The center acts as the CEC's technical advisor on its programs, so assessed and reported on the pilot and demonstration projects, for example.

It has also played a major role in testing solar modules and cells, and in the development of international standards. It participated with JPL[qv], NREL[qv], and others in early "round robins" to standardize test procedures.

Under the management of James Bishop, Tony Sample, and others, JRC had one of the first laboratories in Europe equipped to test PV modules against the emerging standards (Fig. 11.42). Few experiences were more intimidating for the developer of a new solar module than to take prototypes to Ispra, and see

Figure 11.42 Early PV modules at JRC.

guns that would fire ice pellets at your precious samples, the chambers that would cook and freeze them, and the pistons that would deform them! The center also has an outdoor solar test station where modules have been tested, in some cases for decades.

Latterly, JRC has stopped industrial testing of standard solar modules, as several commercial companies do this. It has started to look at operations and maintenance and similar issues, which are increasingly important as the sector matures.

Instituto de Energía Solar at Universidad Politécnica de Madrid

The Technical University of Madrid established its Institute of Solar Energy (IES-UPM) in the late 1970s under the leadership of Prof. Antonio Luque[qv] to work on photovoltaic conversion of solar energy.

The institute has become recognized for its work on solar cell technology using silicon and III–V materials, and new high-efficiency concepts including multijunctions and bifacial cells.

It was responsible for the spinout of the company Isofotón[qv], founded in 1981, initially to commercialize the bifacial cell invented at IES-UPM. Other spin-offs and associated entities include, among others, ETSI-UPM with BP Solar[qv] to apply high-efficiency silicon cells in Euclides concentrators and a research institute devoted to developing large-scale concentrator plants, Instituto de Sistemas Fotovoltaicos de Concentración (ISFOC).

IES-UPM continues to do pioneering research under Prof. Luque, and latterly set a world record in 2008 for GaInP/GaAs cells at a concentration of 1026 suns. The Institute of Solar Energy has latterly broadened its work into areas such as thermophotovoltaics (TPV).

National Renewable Energy Laboratory (NREL)

The US Department of Energy's NREL started life as SERI, the Solar Energy Research Institute, in 1977. It was established by the Carter administration, in Golden, Colorado on federal land previously used by the 10th Mountain Division. The word "Solar" was used broadly to cover various renewables and passive solar. Under its first Director Paul Rappaport[qv], recruited from RCA, SERI's remit was broad including research and dissemination, and it had a budget to match.

In 1980, NREL took over from NASA as the designated national test center for photovoltaic cells (Fig. 11.43). One of Keith Emery's first tasks on joining [171] was to visit Henry Brandhorst at NASA Lewis to "collect the baton." At that time, testing at the module level was mostly undertaken by JPL[qv], so both laboratories participated in the international "round robins" on PV testing described in Section 5.6. On one occasion, the team found an old batch of early 1 in. round silicon cells from Bell Laboratories, probably dating from the 1950s or early 1960s. They measured the efficiency at about 6% – pretty good considering they had no antireflection coating, and the contacts were merely blobs of solder.

Because of its preeminence in independent cell testing, NREL has maintained records on the history of the most efficient laboratory solar cells (see Section 12.1). Its own team under Jerry Olson and Sarah Kurtz set several records for gallium–arsenide solar cells, and successfully licensed patents on this technology. NREL has also held many of the efficiency records for CIGS cells from the late 1990s.

Figure 11.43 Keith Emery at NREL test site in Golden, Colorado.

In the 1980s, NREL's Bhushan Sopori initiated workshops on the "Role of Impurities and Defects in Silicon Device Processing" so that crystalline silicon cell researchers could learn from advances in the wider semiconductor industry.

SERI was central to the generous renewables program under the Carter Administration, undertaking its own research, filing patents, and managing R&D contracts with industry. The incoming Reagan administration had scant interest in renewables and disagreed with SERI's recent publication [172]. Funding was reduced drastically and 300 staff were laid off, including the second executive director Dennis Hayes.

In September 1991, under the administration of George Bush Senior, the center was designated a national laboratory of the US Department of Energy and its name was changed to the National Renewable Energy Laboratory. The center nonetheless continued to experience fluctuating levels of support from successive Washington administrations. Director from 1994, Charlie Gay[qv], again found himself having to slim down the headcount. In-house research activities typically suffered on these occasions, but NREL attempted throughout to remain involved in promising but less popular technologies, especially the compound thin films studied by Tim Coutts and others.

In 1996, the Department of Energy created the National Center for Photovoltaics (NCPV) to be the focal point for developing technology and disseminating information about photovoltaics in the United States. Later led by Larry Kazmerski[qv], it was based at NREL, using their staff, but drawing also on expertise from Sandia and Brookhaven National Laboratories, DOE Regional Experiment Stations, and universities in Georgia and Delaware.

NREL remains a major research and knowledge center for PV and the renewable energy sector in general, and the series of silicon workshops continues.

University of New South Wales

See UNSW-SPREE.

NPAC and Northumbria University

Newcastle Photovoltaics Applications Centre was established by Bob Hill[qv] to become the leading PV research center in the United Kingdom. The solar team at Northumbria Polytechnic College (as it was then) had grown from a handful of people to become the UK's largest academic PV research group by the mid-1980s, when it established NPAC, under the direction of Prof. Hill and managed by Dr. Nicola Pearsall[qv].

The Centre selected some of the less fashionable solar cell materials to work on, majoring on compound semiconductors using abundant and sustainable

Figure 11.44 Northumbria's solar façade [25].

elements. It didn't have the equipment or infrastructure to produce the materials themselves, but sourced the wafers externally and focused more on the properties and performance of the cells than materials science. NPAC has specific expertise in III–V materials, such as gallium arsenide and indium phosphide, primarily for space applications.

An equally important part of its work from the start has been system design, assessment of system performance, integration of PV into building and the electricity network, and environmental impact assessment of the manufacture and use of PV systems.

Northumbria hosted the first major PV façade in the United Kingdom, when it was refurbishing its Ellison Building. This was supported by the industrial partners, most notably Ove Arup and BP Solar, and the UK and European governments. It opened in 1995 and was the UK's participant PV system in the IEA[qv] "solar in buildings" task (Fig. 11.44).

NPAC collaborates widely with industry, has participated in and led many international joint projects, and is a member of Eurec[qv]. Funded by academic and industrial partners across Europe and grants from the UK Government, it has subsequently been renamed the Northumbria Photovoltaic Applications Group.

Solar Energy Research Institute (SERI)

The Solar Energy Research Institute started operations in Golden, Colorado in 1977, pursuant to the Solar Energy Research, Development, and Demonstration Act of 1974. It was later transformed into the National Renewable Energy Laboratory, so see the profile under NREL.

Stanford University

Although Stanford University did not establish a separate PV research institute, it merits mention because of its work on high-efficiency solar cells, its many eminent alumni, and the early support it gave to SunPower Corporation[qv].

Silicon Valley grew up around Stanford University, thanks to the work of its Nobel Prize-winning professor of electrical engineering, William Shockley, and others. Stanford had one of very few integrated circuit fabrication lines at the time under Prof. Jim Meindl, and used professional management rather than academics to run it, so "surprisingly everything worked" [70].

It is natural, therefore, that Stanford has a photovoltaics group working on silicon solar cells. Prof. Dick Swanson[qv] led this work from the late 1970s and throughout the 1980s. One of the main advances was improved efficiency of the contacts that conduct the current from the solar cells. The group developed interdigitated contacts on an insulating layer at the rear of the solar cell, significantly reducing losses associated with shadowing and contact resistance.

Other groups at Stanford have worked on other photovoltaic materials. Prof. Richard Bube had a group focused on cadmium telluride (CdTe) cells and participated in the US Department of Energy[qv] Thin Film Partnership. Bell Laboratories' solar cell pioneer Gerald Pearson was appointed a professor at Stanford, where he worked on gallium arsenide (GaAs) solar cells in the 1970s.

Stanford was one of the first universities to encourage entrepreneurialism in its academics, and established the Stanford Industrial Park to encourage spin-offs. The Silicon Valley magnate Jim Clark was one of the early beneficiaries and started an entrepreneurs club at the University. This approach imbued the support given to Dick Swanson when he established SunPower[qv] and in its early years. When its production was starting up, it was allowed to undertake some process development in Stanford laboratories, while its own facilities were being built.

Several other prominent photovoltaic scientists profiled in Section 10.2 spent time at Stanford, including Roger van Overstraeten[qv], Adolf Goetzberger[qv], and Peter Aschenbrenner[qv].

Stuttgart University (IPE)

The Institute of Physical Electronics (IPE) was the name given by Werner Bloss[qv] to this group of some 15 scientists at Stuttgart University, soon after he took responsibility for it in 1970. Its research was focused primarily on image processing, photovoltaics, and plasma research.

Following the first oil crisis, Bloss persuaded the German Ministry for Research and Technology BMFT to fund a series of research projects for the development of thin film solar cells. Further funding was later secured from the European Commission. The first period of research at IPE worked on copper sulfide/cadmium sulfide thin film cells and modules and led to collaboration with Nukem[qv] and others. At the same time, research at the institute contributed to understanding of the physical properties of amorphous silicon and thin film preparation processes.

From the mid-1980s, IPE continued research on amorphous silicon materials and started work in copper–indium diselenide (CIS), and related compounds, contributing to understanding of material properties and achieving good laboratory efficiencies. The headcount grew steadily to about 50 scientists by the turn of the century.

Latterly, Stuttgart held the efficiency record for the most efficient thin film crystalline silicon solar cell from 2001 until 2011.

UNSW School of Photovoltaic and Renewable Energy Engineering

This group at the University of New South Wales in Sydney (UNSW-SPREE as we'll call it to save trees) became the leading research center for crystalline silicon cells for much of the early PV era. The group dates back to 1974, when Martin Green[qv] joined as a young academic, and grew progressively, starting with Stuart Wenham and Andrew Blakers as its second and third PhD students, respectively.

Their approach was to address each characteristic and element of the solar cell and find ways of optimizing every aspect to enhance cell performance. UNSW-SPREE achieved the world's most efficient crystalline cell, at 18%, in 1984, and held successive records, almost uninterrupted for 30 years. Thus, it was responsible for many of the innovations outlined in Section 4.2. Its second significant break-through was what Green calls "the 4-minute mile of the PV field." Many had felt that 20% efficiency was the upper limit for non-concentrator silicon cells, but UNSW hit it in 1985 [173], and didn't stop there (Fig. 11.45).

The group cannot be accused of ignoring commercial viability in an academic ivory tower; it focused on cost reductions alongside improving efficiency. This led to several industrial licenses of which the first was with AEG-Telefunken, but the most enduring was BP Solar, whose Saturn range of solar modules launched in 1991 owed much to SPREE's technology.

Figure 11.45 UNSW's 20% team – Martin Green at front [25] (they got a bigger lab after that record!).

Later the group also looked beyond crystalline silicon at thin films, which many people expected to overtake crystalline devices owing to their potentially lower costs. In 1995, it initiated a 5-year program supported by Pacific Power to apply its light-trapping technology in thin films, and experiment with a parallel multijunction approach. A spin-off company, Pacific Solar, set up a pilot line for the new technology.

Subsequently, another new company, CSG Solar, was formed to exploit this technology with support from Germany's Q-Cells. UNSW-SPREE later extended its research into what it called "third generation" or heterojunction cells – a sort of hybrid between more efficient "first generation" crystalline technology and lower cost "second generation" thin films.

Many senior technologists in the Far East's leading PV companies were educated at UNSW. Zhengrong Shi, Ted Szpitalak, Jianhua Zhao, and Mohan Narayanan, in addition to Stuart Wenham, have all established production facilities and/or served as Chief Executive or Chief Technology Officers at leading twenty-first century Chinese and Taiwanese producers, including Suntech, JA Solar, Trina, Sunergy, Solarfun, and Yingli.

11.5 International, National, and Representative Bodies

Finally, this section deals with governmental and representative bodies, apart from national laboratories, which have already been covered in Section 11.4.

Commission of the European Community

The CEC (latterly simply referred to as the European Commission) was the governing body of the European Economic Community (EEC – as it was then).

It is divided into Directorates-General, three of which were most involved in the early solar energy sector.

DG-XII Science Research and Development was responsible for most of the early support for the European renewables industry. In 1975, a renewable energy program was initiated under Albert Strub. Two years later, Wolfgang Palz[qv] was appointed to lead the solar work stream, supported at times by Guiliano Grassi and others.

DG-XII supported both research and deployment of photovoltaics through various measures (as detailed in Section 8.4), including the PV pilot projects (Section 13.2). It also implemented the European series of Photovoltaic Conferences[qv] (see Section 14.2).

DG-XII was also responsible for the Joint Research Centre at Ispra in Italy, profiled in the previous chapter.

This directorate subsequently transformed to the Directorate-General for Research and Innovation (DG-RTD).

DG-XVII Directorate-General for Energy was responsible for overall energy policy. Initially, its involvement in the photovoltaics sector was rather sporadic, because solar represented such a small part of the European energy mix.

Most prominent during the early PV era was Dr. Willi Kaut, who had responsibility for the EC's demonstration projects (see Section 13.3). This directorate later initiated more comprehensive programs in the renewable energy sector, including Thermie, under Beatrice Yordi, and Altener.

This directorate subsequently became part of Transport and Energy (DG-TREN), but was split out again in 2010 as DG-ENER.

DG-VIII Directorate-General for Development handled the EC's overseas aid and development. It supported many renewable energy projects in developing countries and initiated specific programs of its own, such as a Regional Solar Energy Programme in the CILSS countries.[3]

This made a significant contribution to the development of pumping, minigrid, and other rural electrification applications for solar energy.

This directorate is now known as DG-DEVCO.

Eurec

Eurec, the Association of European Renewable Energy Research Centres was founded in 1991 as the European Economic Interest Grouping with the goal of improving European research and development in renewable energy.

The purpose of the association is to promote and support development of technology and expertise to accelerate transition to a sustainable energy system.

---

3 See the glossary for a note on what these sub-Saharan African countries are.

Eurec represents prominent research and development groups spread across Europe, operating in all renewable energy technologies, including many of those mentioned in this and the previous chapter and elsewhere in the book.

European Photovoltaic Industry Association (EPIA)

EPIA was launched in 1985 at the London European PV Conference^{qv} by Giovanni Simoni^{qv} and leaders of 16 European solar companies, with strong support from Wolfgang Palz^{qv} at the European Commission. The companies were Ansaldo^{qv}, BMC Solartechnik (from Germany), BP Solar^{qv}, ENE^{qv}, Helios (Italy), Interatom (part of Siemens^{qv}), Intersolar Group^{qv}, Italenergie (Italy, obviously), Photon Technology (Belgium), Photowatt^{qv}, Pragma^{qv}, S International (France), Sama (Greece), and Team (Italy).

The Association was formed to represent industrial PV companies especially in their dealings with the European Commission. A further important role was to enable companies from different countries across the continent to communicate and meet so they could compare notes on national PV programs and share best practice. It also undertook profile raising for the sector in general.

The first board comprised Giovanni Simoni^{qv} as president, Emmanuel Fabre^{qv} and Joachim Benemann ^{qv} as vice presidents, Guy Smekens ^{qv} and Philip Wolfe^{qv}. Guy satisfied the requirement, as a Brussels-registered organization, to have a Belgian national on the board, and he served as treasurer for many years.

After trying for a time to manage the Association with just a subcontracted secretariat in Brussels, the board decided that better executive resource was needed and in 1986 John Bonda^{qv} was appointed as Secretary General. He did wonders to raise the profile of the Association, and was also able to undertake some paid projects in EPIA's name.

The annual general meeting proved to be an important occasion for members to meet and compare notes. Indeed, these meetings became legendary, thanks not least to John's skill at selecting memorable venues. EPIA membership grew steadily to well over 100 by 1999. Because many of the European Commission's PV support programs required collaboration with participants from more than one country, the EPIA proved a fruitful forum for finding suitable partners.

EPIA presidents in our time frame were Giovanni Simoni^{qv} (1985–1987), Emmanuel Fabre^{qv} (1988–1989 and 1994–1995), Eckehard Schmidt of DASA^{qv} (1989–1990), Philip Wolfe^{qv} (1990–1991), Tapio Alvesalo^{qv} (1991–1992 and 1997–1999), Dipesh Shah^{qv} (1992–1993 and 1995–1997), Dr. Dieter Mertig of AEG^{qv} (1993–1994), and Winfried Hoffmann^{qv} (1999 and for a significant share of the subsequent decade). Ibrahim Samak of Engcotec, Stuart Brannigan of BP Solar^{qv}, and others also served on the board (Fig. 11.46).

On John Bonda's death in 1999, Bernard McNelis^{qv} kindly agreed to serve as interim Secretary General until a replacement could be found. He eventually handed over to Murray Cameron about a year later.

Figure 11.46 EPIA 10th anniversary [68]. Directors and former presidents (from left) Dieter Mertig, Ibrahim Samak, and Philip Wolfe.

The Association continues to be an important focal point for the European PV industry, and latterly changed its name to Solar Power Europe in 2015.

Institute of Electrical and Electronic Engineers (IEEE)

The IEEE has a remit very much broader than the photovoltaics sector, but merits brief mention here as the longtime organizers of America's primary series of PV Specialist Conferences further described in Section 14.1.

It was formed in the United States in 1963 from the amalgamation of the American Institute of Electrical Engineers and the Institute of Radio Engineers, and is the world's largest association of technical professionals.

International Energy Agency (IEA)

The International Energy Agency was founded in 1974 in the wake of the first oil crisis to help countries coordinate a collective response to energy security issues. Its remit covers all energy sources apart from nuclear power, which is coordinated by the International Atomic Energy Agency (IAEA).

Although one of the IEA's six founding objectives referred to "alternative energy," much of its early focus was on oil supply issues, and it was for a time seen by the renewable energy sector as a defender of fossil fuel interests. As time passed, sustainable energy sources have formed a larger part of the Agency's work and it has adopted a progressively more supportive role.

The Solar Heating and Cooling Programme (SHC) was one of the IEA's first programs established in 1977 to promote the use of all aspects of solar thermal energy. Although it did not nominally cover PV, its chairman ERDA's Fred Morse was persuaded by Jürgen Schmidt and others to introduce a Task 16 – Photovoltaics in Buildings from 1990 to 1995.

Meanwhile, there was growing pressure, led by ENEL's Roberto Vigotti, to establish a specific photovoltaics work stream. Eventually, the resistance from certain IEA member countries was overcome and the Photovoltaic Power

Systems Programme (PVPS) was established in 1993, with Vigotti serving as the first chairman. See Section 8.8 for further details about the IEA-PVPS.

Latterly, the IEA has accepted that PV will make a substantial contribution to the world's future energy supply. It published its first *PV Technology Roadmap* in 2010, and has since updated it.

International Solar Energy Society (ISES)

ISES is an international membership body for people interested in any form of solar technology – photovoltaics, solar thermal, and passive solar architecture. The society has its origins in 1954 in Phoenix, Arizona, where a group of industrial, financial, and agricultural leaders established a nonprofit organization the Association for Applied Solar Energy (AFASE). The following year it held meetings attracting more than 1000 people from 36 different countries, and the year after that it inaugurated its publication *The sun at work*.

In 1963, it changed its name to The Solar Energy Society and was accredited by the United Nations. One of the cofounders, Prof. Farrington Daniels, was elected president. National chapters of ISES have subsequently grown up in several dozen countries. The first international conference outside the United States was held in 1970, the year when the international headquarters was moved to Melbourne, Australia. "International" was added to the name the following year. The first issue of the ISES magazine *Sun World* was published in 1976. The headquarters was moved again in 1995, this time to Freiburg, Germany.

ISES holds a biennial Solar World Congress on odd years (see Section 14.5) at which it gives various awards (see Section 14.6), including the Farrington Daniels for intellectual leadership and Achievement through Action Awards. The Global Leadership Award in Advancing Solar Energy Policy (catchy title!) has been added latterly in memory of Hermann Scheer[qv].

Ministry of International Trade and Industry – Japan (MITI)

Tsūshō-sangyō-shō, the Ministry of International Trade and Industry (MITI), was an influential department of the Government of Japan, responsible for much of the country's industrial policy. The ministry funded research and directed industrial investment, so was crucial to early Japanese PV businesses.

In 1972, the National Electrotechnical Laboratory proposed adding solar energy research as a new project in MITI's national Large-scale Industrial Technology Programme, which had been running since 1966. Because of Japan's high dependency on imported oil, the ministry was interested in developing new energy sources, but realized that this nascent technology would need longer than the 10-year limit for the existing scheme. MITI, therefore, planned a new R&D program with a longer perspective, adding other energy technologies. At a crucial stage in the appropriation process, the first oil crisis blew up, stimulating greater public and political support for the program.

This led to the establishment of the Sunshine Project in 1974 with a budget of ¥2.5 billion. Initially, solar thermal generation was the top priority, with PV as a secondary technology. This project became the key driver for the Japanese solar energy sector for the last quarter of the twentieth century, as described in Section 8.5.

Following the introduction of Germany's 1000 Roof Programme (see Section 8.4), MITI officials visited Germany and invited BMFT's Walter Sandtner to Tokyo. The Japanese program was updated as the New Sunshine Project in 1994, and bears strong resemblance to the German model.

Subsequently in 2001, MITI's role was taken over by the newly created Ministry of Economy, Trade, and Industry (METI).

National Government Departments from Other Countries

We have not covered governments of individual countries, mentioning only MITI and US DOE because they were responsible for the leading programs on their respective continents, becoming natural counterparts to the pan-national European Commission.

Several other government departments instigated noteworthy national programs of their own, most notably the German activities of BMFT outlined in Section 8.4. The other countries committing significant budgets, as shown in Fig 8.2, were Italy, Switzerland, and the Netherlands.

SEIA, the Solar Energy Industries Association

This US trade association for solar thermal and PV companies has its origin in a meeting of five solar water heater producers at the Washington Hilton in January 1974, convened by Sheldon Butt of Olin Brass. They agreed to set up a trade association, but Butt's concern [101] "don't we need an industry before we can have an industry association?" was shared when they visited a young senate staffer, Scott Sklar[qv]. They, therefore, established a not-for-profit organization, the Solar Energy Research and Education Foundation, which obtained a grant from the Department of Labor to research job potential in the sector.

The trade association was formed in 1980, with Sheldon Butt as its first President; initially with no staff; managed instead by an external service provider. In 1983, SEIA started to take on its own employees, hiring David Gorin as executive director and Scott Sklar as political director. When Gorin left some 18 months later, Sklar was appointed the executive director – a post he held for the next 15 years.

Most of the early pioneering PV companies joined SEIA; the most active being Arco Solar[qv], Spire[qv], Solarex[qv], and ECD[qv]; the latter represented by Vice President Nancy Bacon. The association had an uphill struggle during the early years of the Reagan presidency, needing to be relatively confrontational to prevent all the supportive legislation of the Carter years being entirely swept away. Despite the political climate, it was able to promote ongoing research and development funding, keeping solar energy a priority for the Department of Energy. By the 1990s, it had a membership of 150 companies, many by this time being from the photovoltaic sector.

Many PV members served on the board of SEIA. Presidents from the sector included Roger Little[qv], Walter Hesse of Entech (in the 1980s), Allen Barnett[qv] (1995–1997), Solarex's Harvey Forest (1997–1998), and Mike Davis from Kyocera[qv].

Scott Sklar left SEIA in 1999, but returned for a 6-month period, when his successor left. Subsequently, Rhone Resch was appointed executive director, serving for 13 years. The association has latterly continued to grow, to over one thousand members, and remains a prominent promoter of US solar industry.

US Department of Energy

The US Department of Energy (US DOE) was established in 1977 during the Carter Presidency, consolidating the activities of predecessor agencies, including the Energy Research and Development Administration (ERDA), the Federal Energy Administration, the Federal Power Commission, and Nuclear Regulatory Commission. The first Secretary was James Schlesinger. An important part of its remit during this administration, with the first oil crisis fresh in the memory, was support for energy conservation and renewable energy. These were called "solar" for convenience, but embraced a broader range of technologies.

Also in 1977, the Solar Energy Research Institute (SERI[qv]) was established in Golden, Colorado with Dr. Paul Rappaport[qv] as its first director. An Assistant Secretary for Conservation and Solar Energy was appointed and he established a Photovoltaic Energy Systems Division, headed up by Paul Maycock[qv], supported by Morton Prince[qv], and later by Bud Annan. This division managed the US government's program to support the US photovoltaic sector, further detailed in Section 8.3. As described there, the sector had a roller-coaster ride of fluctuating fortunes, depending on the political preferences of successive White House occupants. It also encouraged the use of PV by other federal departments.

The US DOE has a number of national laboratories. While SERI, which in 1991 became NREL, the National Renewable Energy Laboratory[qv], is specifically tasked to lead in renewables work, other US DOE labs, notably Sandia and Brookhaven, have also played a part. Interdepartmental support also comes from Jet Propulsion Laboratory, which is part of NASA.

At the time of writing,[4] US DOE continues to have a division tasked to support renewable energy under the Under Secretary for Science and Energy.

World Bank Group

The World Bank Group evolved from the International Bank for Reconstruction and Development (IBRD) founded in 1944 as facilitator of postwar reconstruction by lending to national governments. The group also includes the International Finance Corporation (IFC), which lends to the private sector. Its remit has evolved over time to focus on worldwide poverty alleviation, often in partnership with the Global Environment Fund (GEF).

In 1989, the Bank, working with UNDP and others, initiated a study on "Financing Energy Services for Small Scale Energy users" (Finesse). This led to the establishment in 1992 of the Asia Sustainable and Alternative Energy Program (ASTAE) under Loretta Schaeffer to stimulate small-scale sustainable energy projects in Asia. The initial phase was funded by grants approved by Bud Annan of US DOE[qv] and Paul Hassing of the Netherlands overseas aid department DGIS.

An early recruit was Anil Cabraal, soon followed by Mac Cosgrove-Davies. The unit went on to support PV, and other sustainable energy, projects in many countries including Sri Lanka, India, China, Indonesia, the Philippines, and Bangladesh (Fig. 11.47). The team not only sponsored projects but also published papers on the best practice lessons learned [174] and the benefits of deploying PV in such regions [175].

Figure 11.47 Rural solar power in Bangladesh supported by World Bank [68].

---

4 Early in the Trump presidency.

Anil Cabraal went on to become manager of the Lighting Africa program, served on the board of PV-GAP, and was instrumental in designing and launching the small-scale renewable energy program. He subsequently received the Prof. Robert Hill Award for the promotion of photovoltaics for development. His chapter in the *Solar Power* book [43] gives further details of World Bank's involvement in PV, as does that of Bernard McNelis^{qv}.

A later much-heralded initiative of the World Bank's IFC and GEF was the PV Market Transformation Initiative (PVMTI or "Green Carrot"). This was launched in 1998 to stimulate a major transformation of the emerging PV market by accelerating commercialization and financial viability of PV technology for rural electrification in the developing world. Impax and IT Power^{qv} were selected to manage the program.

The World Bank Group continues to support PV and other renewable energy projects, including utility-scale generating stations.

11.6 Not Individually Profiled

Because of the selection criteria detailed at the start of this chapter, some notable organizations do not appear in the list above, including the developers of the first viable solar cell, Bell Laboratories.

Companies in PV mainly for space have, with one exception, not been profiled; notwithstanding major contributions to the early and wider PV industry by Applied Solar Energy Corporation[5], Hoffman Electronics, Martin Marietta, Messerschmitt-Bölkow-Blohm (MBB)[6], NASA, the UK's Royal Aircraft Establishment, TRW, and Varian Semiconductor. Nor has Boeing been directly involved in terrestrial PV, although it has made an indirect contribution by setting several of the early records for efficiency in CIGS solar cells.

Some of the pioneering research (see Chapter 3) during the early PV era on all sorts of novel solar cell materials and processes never made it into volume commercial production. Active companies included Fred Schmid's Crystal Systems, Mobil Tyco[7], Motorola, Texas Instruments, and Westinghouse in the United States, Philips[8] in Europe, and Hitachi, Toshiba, and NEC in Japan.

Significant producers, who operated primarily in their home market of India were Central Electronics Limited (CEL) and Bharat Heavy Electricals Limited (BHEL). Japanese producers Hoxan, Taiyo Yuden, and Kaneka could all

---

5 This was also involved in development of concentrator cells.
6 See ASE^{qv} and Phototronics and Solems under Total^{qv} in Section 11.2.
7 See ASE^{qv} in Section 11.2.
8 See Photowatt^{qv} in Section 11.2.

arguably have been included, having reached the top 10 in the 1980s, before pulling out in the 1990s. Some companies had terrestrial photovoltaics activities, which never emerged as significant business streams. In the United States, these included Optical Coatings Laboratory (OCLI) and solar concentrator company Entech. Similarly, in Europe, Ansaldo, Ferranti, and Pilkington[9] briefly had PV activities.

There were many smaller companies that, with varying degrees of success, conducted business in our time frame. Second tier producers included Franco Traverso's Helios in Italy, Bernd Melchior and Oussama Chehab's BMC Solartechnik in Germany, and in the United States Irwin Rubin's Sensor Technology[8], and Solenergy founded by Bob Willis[qv]. Systems integrators, distributors, and other downstream companies are too numerous to name, although some were quite sizeable, such as Olaf Fleck's Sunset Energietechnik and Phoenix-based Photocomm, headed at various times by Arco Solar alumni Bob Kaufman and Tom Dyer, and eventually bought by Kyocera[qv].

In addition, there are those, who started out in the last millennium, but went on to make their major contribution after our time frame – to be properly recognized, hopefully, should someone write the sequel to this book. Again, the list is long and includes the pioneers in utility-scale solar profiled in my previous book [44]. Other obvious candidates in Europe include Antec Solar, PV Silicon[10], Q-Cells, Solar Fabrik, and SolarWorld. In the United States there is EPV, First Solar and its predecessor companies set up by Hal McMaster with Arun Madan, Glasstech Solar, and Solar Cells Inc. Japan's Fuji Electric started thin film research in 1978, but didn't go into mass production until 2006. Many of today's leading producers started after our time frame, including the major Chinese producers, which latterly became so dominant.

In the supply chain, Applied Materials' business dates back to the 1960s, but its prominent position in the supply of PV manufacturing equipment started shortly after our time frame.

11.7 Major PV Producers, Entrants, Floaters and Leavers

We end this chapter with a summary of leading producers, and of major entrants and leavers, together with company flotations during our time frame (Table 11.1).

---

9 See Flachglas[qv] in Section 11.3 and Shell[qv] in Section 11.2.
10 See Crystalox[qv] in Section 11.3.

Table 11.1 Major PV producers, market entrants, and departures.

	Top producer	Other top 5	Entrants[a]	Mergers, *IPOs*[b]	Leavers[a]
Pre-1975	Sharp Philips RTC		Sharp, Philips AEG, RCA Shell, Exxon		
1975-1979	SPC (Solar Power Corporation) Solarex	Sharp Arco Solar Sanyo Photowatt	Kyocera, Sanyo Arco RWE (Nukem) Amoco		
1980-1984	Arco Solar	Sharp Solarex Fuji AEG	Siemens ENI Neste Flachglas Texas Inst.	BP Lucas Photowatt *Spire*	RCA, Exxon
1985-1989	Sanyo[c] (thin film)	Fuji Solarex Hoxan Photowatt BP Solar	ENEL	DASA *ECD*	
1990-1994	Siemens Solar	Sharp Kyocera Solarex BP Solar Solec		ASE	Arco
1995-1999	Kyocera BP Solarex	Sharp Solec AstroPower ASE Sanyo	ATS	BP Solarex *AstroPower* *SolarWorld*	Texas Inst. Amoco
Later	Sharp		EDF	*SunPower* *Crystalox, SMA*	BP, Neste, ATS Flachglas

This list is based on information sources listed at the start of section 5. The independent sources of market information at the time were not all consistent and the data may be incomplete, so I apologise for any inaccuracies or omissions in the table.

a) Showing only multinational companies with significant activities in the sector.
b) IPOs are shown in italics, consolidations in standard text.
c) By some measures Sanyo's thin film sales made it the leading volume producer in 1986–1987.

12

How: Research and Technology

This brief chapter looks at the efficiency of solar cells, both theoretical and actual, and the technical details of the standards against which they are tested.

12.1 Cell Efficiency Trends and Records

There is no better reference to show the progress of PV cell efficiency achieved in the laboratory than the chart produced by the National Renewable Energy Laboratory [176]. Fig. 12.1 shows an extract from this chart covering our time frame. There has been plenty of progress since and the right-hand side (not shown here) displays continuing increases in efficiency and several new materials.

There are a few "health warnings" to remember, when studying this chart. First, it only includes independently validated results. Some higher efficiencies have been claimed, but are not included if NREL[qv] couldn't verify them. The efficiencies are from individual laboratory cells, so represent the best achievable at the time, not what could be reproducibly manufactured on the production line. The areas of some of these test cells are very small and may not therefore be representative of how full size cells would perform because of resistance and "edge effects."

NREL's Keith Emery also points out [171] that in comparing the records for different materials, these results will tend to flatter concentrator cells (shown with dashed lines), because they are measured only against direct radiation. Other cells types will also, in practice, benefit from the diffuse element of radiation, but concentrators won't.

The darkest lines on the chart are for mono- and polycrystalline silicon cells and illustrate the sequential records achieved by Martin Green's group at UNSW-SPREE[qv], with a brief interruption by Dick Swanson[qv], who also held the silicon concentrator records with the Stanford[qv] and SunPower[qv] teams.

The Solar Generation: Childhood and Adolescence of Terrestrial Photovoltaics, First Edition. Philip R. Wolfe.
© 2018 by the Institute of Electrical and Electronic Engineers, Inc. Published 2018 by John Wiley & Sons, Inc.

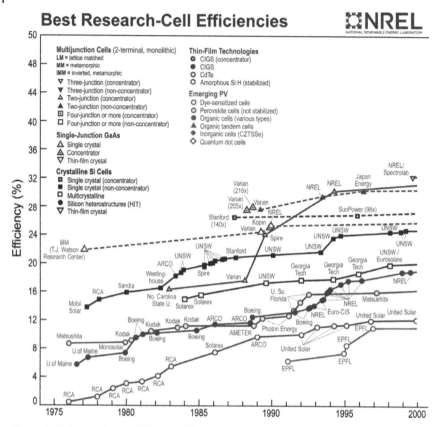

Figure 12.1 Extract from NREL's cell efficiency progress chart [176]. (Courtesy of the National Renewable Energy Laboratory, Golden, CO.)

Nonconcentrator monocrystalline efficiency increased from 14 to 25% during the period, and polycrystalline reached almost 20%.

Thin film amorphous silicon records (the lighter color second from bottom) were set by the RCA[qv] team, moving with them to Solarex[qv], until Arco Solar[qv] set a record in the late 1980s and then ECD[qv], through its United Solar subsidiary, set several sequential records. The two lines above this are for thin film copper–indium–gallium diselenide (CIS) and cadmium telluride (CdTe), with records spread between more different participants, until NREL themselves set a series of CIS records in the mid- and late 1990s.

All records during the period for dye-sensitized cells (the lightest color at the bottom) were set by EPFL[qv]. The records for gallium arsenide and other space cells are also shown at the top.

12.2 Theoretical Efficiency Limits

A solar cell's energy conversion efficiency is the percentage of power converted from sunlight to electrical energy under standard test conditions (STC see further).

William Shockley and Hans Queisser calculated in 1961 [177] that the maximum efficiency of a single-junction silicon solar cell is about 30%. This became known as the Shockley–Queisser efficiency limit (SQ Limit). Later, refinements to the calculations show that the efficiency limit is about 33% for any type of single-junction solar cell.

Higher efficiency limits apply to multijunction cells. In fact, it has been calculated that a cell with a very large number of junctions could have a limiting efficiency up to 86.8% under highly concentrated sunlight [57].

There are some assumptions in calculations of the SQ Limit that restrict its general applicability to all types of solar cells, but it can be accepted as a good approximation to the level achievable for the types produced in the early PV era.

12.3 International Standards for Cell and Module Testing

International collaboration in the development of test standards for solar cells and modules was outlined in Section 5.6.

12.3.1 Norms for Rating and Testing PV Devices

The following norms apply to the testing and qualification of solar cells, modules, and systems:

Air mass (AM) (see Glossary)	PV standards use AM1.5 equivalent to the sun at an angle of about 48° from the zenith. This leads to a standard for the spectral irradiance distribution used during testing.
Nominal operating cell temperature (NOCT)	The steady-state temperature of a cell in a solar module with uncovered rear surface, when exposed to insolation intensity of $0.8\,\mathrm{kW/m^2}$ normal to the plane of the solar module, at ambient air temperature of 20 °C with wind velocity 1 m/s
Standard test conditions (STC)	For solar module testing, defined in international standards; insolation intensity of $1\,\mathrm{kW/m^2}$ normal to the plane of the solar module at ambient temperature of 25 °C and light spectrum equivalent to AM1.5
1 sun	Shorthand nomenclature for conditions equivalent to full sunshine as defined under STC.

12.3.2 Solar Module Qualification Testing

To be accredited against the international standards referred to in Section 5.6, solar modules are rated in terms of watts peak under standard test conditions (see above). They are then tested to ensure they are adequately strong, safe, and weatherproof, with a series of tests (also conducted against norms defined in the standards).

The following tests are typically performed and the results are used to validate ratings given by the manufacturer:

Visual inspection; test performance at STC; insulation test; measure temperature coefficients of open-circuit voltage V_{OC} and short circuit current I_{SC}; measure NOCT (see above); test performance at NOCT; test performance at low irradiance; outdoor exposure; hot spot endurance test; UV exposure; thermal cycling; humidity and freeze cycles; damp heat; test terminations; twist test; mechanical load test; and hail test.

For thin film modules, the following additional tests are carried out: light soak; thermal annealing; and wet leakage current.

13

Where: Geography and Politics

This chapter provides specifics of some of the key policies outlined in Chapter 8.

13.1 US Programs and Cost Targets

US policy, as described in Section 8.3, had a strong focus on working with industry to bring costs down and thereby expand the markets for PV. This section gives some of the specifics of these targets and activities.

13.1.1 Cherry Hill Conference Key Program Targets

This 1973 conference [59, 103, 178] sought to agree a program of work for the next decade, and identified targets and the anticipated investment to reach them (Table 13.1).

The Cu_2S/CdS expectations proved unduly optimistic, as mentioned in Section 4.5, but otherwise, this action list highlighted appropriate fields to work in and set the background against which the early industry progressed.

Table 13.1 Cherry Hill Conference key program targets.

Development field	Cell cost target ($/W_P)	Production volume target (MW/year)	Development investment required ($million)
Monocrystalline silicon	0.50	500	450
Copper sulfide/ cadmium sulfide	0.20	1000	183
Polycrystalline silicon	0.50	10	45
Other device materials	Identify 1 new type		14
Testing and evaluation	Test laboratory		9
PV systems	100 kW systems		15

The Solar Generation: Childhood and Adolescence of Terrestrial Photovoltaics, First Edition. Philip R. Wolfe.
© 2018 by the Institute of Electrical and Electronic Engineers, Inc. Published 2018 by John Wiley & Sons, Inc.

13.1.2 JPL Block-Buys

The Jet Propulsion Laboratory[qv] undertook a series of bulk purchases of PV modules from selected manufacturers starting in 1975. A core purpose of the Block-buys was to develop test standards and help producers improve quality.

For Block I, the specification was simple, it required verifying the electrical test per manufacturers' ratings and carrying out environmental tests limited to temperature cycle and humidity soak. Even these simple tests required some design improvements during production. In Block II, the design requirements were extended to include interconnect and terminal redundancy. Structural loading was added to the tests. A quality management requirement was also included in Block II, which predates the first ISO quality management system in 1987.

The specifications were developed further in subsequent blocks to reflect failures and to make them better suited to terrestrial applications – JPL's original specs having been tailored to space conditions.

These Block-buys proved invaluable to suppliers in improving design and product quality. A dramatic reduction in module infant mortality occurred between Blocks IV and V. The Block V test specification became the basis for IEC's international standards, as described in Section 5.6. The rounds were as outlined in Table 13.2.

An equally important aim of the Block-buys was to provide some volume throughput for the US photovoltaics industry, which was significant enough to help cost reduction – 200 kW$_P$ was a substantial volume in 1978. In the later blocks, the volumes were smaller but successful participants became eligible to supply other federal contracts.

Table 13.2 JPL Block-buys: Timing, volume, and suppliers.

Block and year	Volume (kW$_P$)	Producers involved
Block I 1975–1976	58	M7 International, Sensor Technology, Solarex, Solar Power Corporation, Spectrolab
Block II 1977	123	Sensor Technology, Solarex, Solar Power Corporation, Spectrolab
Block III 1978	205	Arco Solar, Motorola, Sensor Technology, Solarex, Solar Power Corporation
Block IV 1980	32	Arco Solar, Applied Solar Energy Corporation, General Electric, Motorola, Photowatt, Solarex, Spire Corporation
Block V 1983	9	Arco Solar, General Electric, Mobil Solar, Solarex, Spire Corporation
Block V Later procurement		Applied Solar Energy Corporation, Motorola, Solavolt, Solenergy

Table 13.3 US National Photovoltaics Program price goals (1980 dollars).

Year	Module price ($/W$_P$)	System price ($/W$_P$)	Energy price (¢/kWh)	Prime application
1982	2.80	6.0–13.0	50.0–90.0	Remote/rural
1986	0.70	1.6–2.2	5.2–8.7	Residential
		1.6–2.6	5.5–9.2	Intermediate
1990	0.15–0.40	1.1–1.8	4.2–8.1	Utility

13.1.3 Department of Energy Program

The overall top level objective of the program as stated [90] in 1980 was "to replace annually by 2000 fossil fuels equivalent to 1 Quad". 1 Quad is 10^{15} BTU, about 300,000 GWh or 25 m tonnes of oil.

US DOE set module and system price targets, expressed in 1980 dollars, as listed in Table 13.3 and illustrated in Fig. 13.1, which also shows (by the square

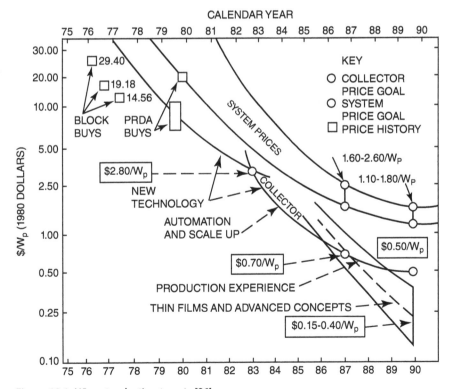

Figure 13.1 US cost reduction targets [86].

Table 13.4 US National PV Program subprograms and work streams.

Subprogram	Work streams	Manager
Advanced cell research	Polycrystalline silicon Amorphous silicon Cadmium sulfide	Solar Energy Research Institute
Long-range research	Gallium arsenide Advanced concentrators Emerging materials Electrochemical cells Innovative concepts	Solar Energy Research Institute
Low-cost solar arrays	Silicon material and sheet Processes and equipment Encapsulation Modules and arrays Qualification and testing	Jet Propulsion Laboratory
System engineering	Balance-of-Systems Systems applications Concentrator technology	Sandia Laboratories
Stand-alone applications	Pilot projects	NASA-Lewis Research Center
Residential applications	Pilot projects	MIT Lincoln Laboratory

boxes at the top left) the prices achieved in the JPL Block-buys. The respective markets accessible at each price point are also identified.

The program had, at the time, the work streams listed in Table 13.4:

With effect from 1990, this became the PV Materials and Technology Program (PVMaT). Its objectives were defined more specifically "to help the US PV industry improve photovoltaic manufacturing processes and equipment; accelerate manufacturing cost reductions for PV modules, balance-of-systems components, and integrated systems; increase commercial product perform-ance and reliability; and enhance the investment opportunities for substantial scale-ups of US-based PV manufacturing plant capacities."

As these programs evolved, new work streams were added. One such initiative, launched in the 1990s, was Photovoltaics for Utility Scale Applica-tions (PVUSA). This supported Pacific Gas and Electric's 500 kW Kerman project, among others, and laid the groundwork for utility-scale applications, which subsequently became widespread in the next century.

13.2 European Pilot Projects

The pilot projects supported by the DG-XII programme described in Section 8.4 are listed further (Table 13.5).

Table 13.5 First European Pilot Projects.

Application	Country	Location	kWp
Vacation center	Germany	Island of Pellworm	300
Village	Denmark	Vester Bögebjerg near Korsoer	100
Village	Greece	Kythnos Island	100
Island with 120 inhabitants	Italy	Alicudi Island	80
Power supply to the grid	The United Kingdom	Marchwood (later moved to Scottish island)	80
Water desalination	Italy	Tremiti Islands	65
Power for swimming pool	Belgium	Chevetogne, Namur	63
FM transmitter	France	Montpellier	50
Nice airport	France	Nice	50
Dairy farm	Ireland	Fota Island, near Cork	50
Marine training school	The Netherlands	Terschelling Island	50
Village	Greece	Crete Island	50
Water disinfection, ice-making for agricultural cold store	Italy	Giglio Island	45
Dwellings, dairy farm and workshop, plus water pumping	France	Rondulinu Cargese, Corsica	44
Village	French Guyana		35
Hydrogen production by electrolysis for factory	Belgium	Hoboken, Antwerpen	30
6 × 5 kW systems compared to thermodynamic solar system	Italy	Adrano, next to "Eurelios" plant	30

Copies of the final report [179] of this program are scarce, but if you can find one, they give a good summary of each project, with photographs.

13.3 European Demonstration Projects

The European Commission's energy demonstration projects supported about 100 PV projects between 1978 and 1991. These are classified according to the categories listed in Table 13.6.

Table 13.6 Application categories of European demonstration projects.

Category	Description
1	Grid-connected houses
2	Farms and hostels
3	Island community houses
4	Remote houses
5	Nature reserves and recreation
6	Lighthouses and light buoys
7	Water pumping and treatment
8	Warning and monitoring systems
9	Telecommunication and public lighting

The three most supported categories were remote houses, islands, and water treatment. Three countries account for the lion's share of the projects: France, Italy, and Spain each handling about 25% of the total capacity.

Many of these installations survive to this day. The Gavdos Island project, for example, can still be seen on satellite images of Europe's most southerly landmass (Fig. 13.2). The project brought power to this previously unelectrified island, and was one of the first to use seasonally adjustable array mounting structures.

The following projects were supported (Table 13.7). The first three were before 1983 and the horizontal lines after that mark the years when the projects were supported, ending in 1990.

Figure 13.2 The Gavdos project still visible in 2017 [45].

Table 13.7 PV projects under European demonstration program.

Category	Application	Country	Location	kWp
	PV powered electric car	Belgium	Leuven	1.8
7	Microirrigation on farm	France	Bourriot-Dourgonce	0.9
	Interconnection with LV grid	Belgium	Gent	5.0
8	Fire prevention + high water alarm	Italy	L'Aquila and Pescara	0.4
2	Cogeneration for farmhouse	Germany	Sachsenheim	5.0
7	Irrigation pumping	Greece	Karpathos Is.	10.0
3	Two remote villages	Greece	GavdosIs.	20.0
3	Village + isolated houses	Greece	Antikythira Is.	35.0
3	PV station for remote village	Greece	Arki Is.	25.0
4	40 isolated houses	France	Southern France	32.0
8	Ground water measurement	Germany	Hamburg	0.3
9	Isolated radio-relay station	Greece	Antikythira Is.	20.0
2	PV electricity for Oberlin House	Germany and France	Baden-Wurttemberg	2.6
5	Bird + weather station	Germany	Scharhorn	4.1
1	PV/Wind for 9 on-grid houses	UK	Milton Keynes	4.1
1	Grid-connected house	Ireland	Farnanes, Cork	5.0
2	Equestrian center at Prado Rondanino	Italy	Ligurian Appenines	2.1
6	Unmanned Lithari lighthouse	Greece	Skyros Is.	2.6
6	PV/wind power for lighthouse	Greece	Sapientza	2.6
6	PV supply for lighthouse	Italy	Palmaiola Is.	5.1
5	Lighting + water pumping for archaeological sites	Italy	Cetano and Sorano	26.1
4	Alpine lodgings + huts	France	French Alps	5.3
7	Stand-alone PV/wind for city park	Germany	Freiburg im Breisgau	2.1
5	La Palissade nature reserve	France	Camargue	12.0
9	PV/diesel for microwave relay	France	Gros Bessillon	1.9
9	Kaw Mountain TV+FM site	France	French-Guyana	20.0

(*continued*)

Table 13.7 *(Continued)*

Category	Application	Country	Location	kWp
4	30 isolated houses in Ginostra	Italy	Eolian Is.	9.0
7	Desalination, refrigeration + lighting	Greece	Helos Is.	31.9
2	Dairy farm on the mountains	Italy	Abruzzo	6.3
8	Highway fog detection	Italy	Verona	14.3
3	Autonomous housing complex	Greece	Karpathos Is.	4.2
1	Grid-connected house	Germany	Ensheim	8.0
1	Electricity for OKAL house	Germany	Berlin	5.1
4	PV/thermal mountain refuges	France	Alpes, Pyrenees	8.4
2	7 dairy farms in mountains	Italy	Liguria	27.4
6	Punta Libeccio light-house	Italy	Sicilia	18.2
7	Water pumping in rural area	France	Corse	3.8
4	Houses, TV repeater, beacon, water treatment	France	Corse	9.8
9	3 telephone exchanges	France	Guadeloupe	12.4
4	Isolated rural dwellings	Portugal	Portugal	35.0
7	Reverse osmosis plant	Spain	Almeria	23.1
6	Light buoys at different latitudes	France	Atlantic, Mediterranean, Fr. Antilles	24.0
3	Island power plant	Spain	Tabarca Is.	100.0
4	57 dwellings	Spain	Sierra de Segura	28.2
4	35 rural houses	Spain	Oden	18.9
8	Airport signaling lights	Italy	Lucca - Tassignano	4.8
2	Passive thermal - PV hybrid	Italy	Palonbara Sabina	3.0
9	PV equipment at international activity park	France	Valbonne - Sophia Antipolis	8.3
8	Insect monitoring with α-Si	France	Montpelier	10.0
9	PV equipment in Roya Valley	France	Alpes Maritimes	6.1
9	Solar energy + tourist trade	France	Cap Corse	8.8
5	Education center Puerto Penas	Spain	Huelva	18.0
2	Agricultural + cattle operation	Spain	Badajoz	12.9

Table 13.7 (*Continued*)

Category	Application	Country	Location	kWp
7	Biological water treatment	Spain	Madrid	12.5
5	National cycling museum	France	Les Ecrins	3.4
9	Lighting, DAS + alarm system for tourist paths	Italy	Riomaggiore, Liguria	15.4
3	Santo Stefano Island – distributed PV systems	Italy	Isola Santo Stefano	21.0
8	Railway to Saline di Volterra	Italy	Cecina, Toscana	17.3
5	Mont Chiran site	France	Haute Provence	1.4
7	Reverse osmosis desalination	Italy	Filicudi Is.	63.0
4	PV station for remote village	Spain	Rambla d. Banco	11.8
4	Power plant for small village	Spain	Fuente Alamo	10.0
2	Agricultural applications	Spain	Alpuharra	18.4
2	Water heating for goat stable	Spain	Malaga	4.3
2	Farm houses (radio-control)	Spain	Granada	12.3
1	Commercial building with α-Si	UK	Cardiff Bay	20.4
7	Power + water purification	Italy	Ventotene Is.	32.0
2	Refrigeration for cheese dairy	France	Briancon	8.6
8	Railway system near Siena	Italy	Honte Antico	12.0
9	Wind/PV telephone exchanges	France	Martinique Is.	8.8
9	PV for bus shelters	France	Chalon s. Safine	5.5
5	PV/thermal for nature park	France	Vercors	10.1
4	Remote district development	France	Corse	5.4
8	Corrosion protection + pipeline monitoring	Germany	Schleswig Holstein	13.5
4	PV system for remote houses	Germany & Italy	Dorneck, Knoepflesbrunn, Bognago	10.4
3	Power plant in Porto Santo	Portugal	Porto Santo	100.0
4	PV/wind for 22 rural houses	Spain	Alt Urgell	11.5

(*continued*)

Table 13.7 (*Continued*)

Category	Application	Country	Location	kWp
4	Remote houses, water treatment + pumping	France	Haut Jura	17.6
4	Isolated PV systems managed by Argos system	France	French & Spanish Pyranees	30.5
4	PV/diesel for houseboats + barges	The Netherlands	The Netherlands	8.8
4	PV for remote sites in Alps	Germany	Bavaria	8.8
9	RAI-TV + radio station	Italy	Pratomagno	14.0
9	Hybrid PV for radio station	France	Mont St Sauveur	13.3
5	Appennino tunnel PV system	Italy	Appenines	10.0
4	Houses in Stromboli + Alicudi	Italy	Eolian Is.	24.5
7	Liquid Manure Storage	Germany	Schleswig-Holstein	2.8
4	Rural Electrification	Spain	Andalucia	12.0
4	Mountain refuges	France	Rhone-Alpes	3.1
7	St Victoire mountains fire protection	France	Bouches du Rhone	27.9
8	Aircraft landing systems	Italy	Toscana	20.6
4	Rural electrification	Spain	Soria	21.2
2	PV/Wind/solar thermal farms	Spain	Catalonia	15.1
7	Combined Electrodialysis	Germany	Tenerife	32.0
4	Rural Electrification	France	Alpes Haritines	25.3
3	Island power + water supply	France	Glenan Is.	19.3
5	Rural electrification for protected nature areas	France	France and Overseas	11.8

13.4 Feed-in Tariffs and National Policies in Europe

Germany is often credited as the inventor of feed-in tariffs, but this is not actually the case. In fact, the concept was first adopted in 1991 in the Swiss town of Burgdorf, thanks to Prof. Heinrich Häberlin and the chief of the local utility, Theo Blättler. However, it was Germany, led initially by the town of Aachen, which made this approach widespread.

The principle of feed-in tariffs is that renewable electricity generators are guaranteed a fixed price for every kilowatt-hour (kWh) of electricity they produce over a period of 20–25 years. This price is paid to them by the electricity utility to which they are connected. The utility, in turn, is entitled to pass on any extra costs it incurs under the scheme to all of its customers. In this way, the cost of renewables is passed on uniformly to all energy users.

This offers government the huge benefit that it doesn't need to finance any part of the installations, as it would under a system of grants, for example. The capital cost is borne by the householder or private investor, and the fact that they will receive a known tariff for many years provides the "bankability" for them to obtain loans against this cost, if they want.

13.4.1 Timeline of Feed-in Tariffs in Germany

The 1000 roof programme, introduced in 1990 by the research ministry BMFT, laid some of the foundations for the feed-in tariff, even though it was itself based on capital grants.

The Stromeinspeisungsgesetz (Electricity Feed-in Act) enacted the next year, gave renewable energy producers, including those with solar panels on their roofs, the right to connect to the grid. This act also introduced the concept of prescribed tariffs for energy sold back into the grid. However, the terms in which these tariffs were expressed were not popular with the utilities; nor did they give sufficient certainty to raise external funding for the solar system.

In Aachen, the renewable energy enthusiast Wolf von Fabeck, who had followed developments across the border in Burgdorf, recognized that a better method of calculating the tariffs was required, to give a viable return to the solar electricity generator. He did his own computations and came up with a proposed tariff of 2 Deutschmarks per kilowatt-hour, equivalent to just over €1 and about 20 times the price of electricity prevailing in Germany at the time. Although this sounded high, von Fabeck also calculated how this tariff would impact on the wider community. Because the cost is spread between all energy users, he showed that the overall effect on consumers would be very small.

With strong support from the good citizens of Aachen, von Fabeck was able to persuade the City Council to vote in favor of the introduction of a feed-in tariff at this level in September 1992. This became known as the "Aachen model," and was soon adopted in other municipalities around Germany, most notably in Bavaria.

The local electricity company, Stawag (Stadtwerke Aachen AG), was less enthusiastic, even though it is the city's municipal utility. It was not opposed to solar as such, indeed it had already been a pathfinder in installing a solar plus solar façade from Flachglas (see page 236), but it objected to the level of the tariff and the extra administration involved. Extended negotiations over 3 years, and independent reconfirmation of the tariff levels were needed before they were implemented in Aachen, by which time several other municipalities had already adopted to the "Aachen model" ahead of them.

Meanwhile, Hans-Josef Fell^{qv}, a member of the federal parliament from the Green Party, had been tracking these developments and was keen to introduce feed-in tariffs nationwide. At the same time, Hermann Scheer^{qv} of the Social Democratic Party was promoting the concept of a 100,000 roofs programme as a follow-on from the successful 1000 roofs programme.

Fell persuaded Scheer that his ambitious project would not succeed solely as a grant-based program because of the capital cost involved, but that it could flourish under a system of feed-in tariffs. They formed an informal partnership and eventually managed to persuade their respective parties and in due course the governing coalition and the German parliament to introduce the Erneuer-bare-Energien-Gesetz (EEG – Renewable Energy Act). This was approved in 2000 leading to an explosion in deployment, particularly after the 2004 revision when tariffs were revised and caps removed. The rest, as they say, is history.

Refer to Bob Johnstone's book [82] for a more comprehensive and enter-taining story of the introduction of feed-in tariffs.

13.4.2 Timeline of Skepticism in the United Kingdom

In stark contrast, the United Kingdom, which enjoys a sunshine regime fairly similar to that in Germany, maintained a stolid skepticism about the relevance of PV and most other renewables; although there were a few fluctuations along the way.

In 1976, the government forecast in Energy Paper 11 [181] that by 2025 nuclear power would deliver 65% of all UK primary energy.[1] "Renewable sources would collectively contribute less than [4 percent] by 2000[2] . . . The two technologies which show most promise would be wave power and solar heat."[3] The following year Energy Paper 16 [81] specifically assessed PV concluding that "the present manufacturing costs are too high by a factor of 1000" and "even if collector costs are eventually reduced photovoltaic devices cannot be considered seriously as a prime source of electricity in the United Kingdom either for replacing or contributing to power in the national grid."

No specific support was offered to the country's early PV companies, even after BP entered the industry in the early 1980s. The Department of Trade and Industry (DTI) had an Electrical Technology Committee, administered by Joe Heinson, which provided R&D funding in the electrical engineering sector, and it did support a handful of PV projects. Similarly, the Department of Energy's alternative energy section under Godfrey Bevan found ways of working "under the radar" to help companies participate in EC projects and to provide marginal support.

---

1 Now, with the benefit of hindsight we know this is unlikely to exceed 20% of electricity.
2 Renewables did contribute about 4% in 2000, and this had risen to some 25% by 2015.
3 These two technologies remained almost entirely unexploited in 2000 and a decade later.

Looking at the sector again in 1994, Energy Paper 62 [182] calculated that "even with optimistic assumptions . . . the cost of electricity from PV in the UK is likely to be about 10p/kWh⁴ (1991 prices) in 2025." However, it conceded that "there is a major world-wide effort . . . A DTI programme would enable the UK to benefit from this investment at only a small cost." A small program was initiated, whose funding peaked at £2.1 million in 1994–1995 then fell back again to £1.6 million in 1997–1998 [183].

Even after our time frame the UK government took time to "see the light." Its 2009 Renewables Strategy [184] projected that all "small renewables" (which include solar) would contribute 4GW by 2020.⁵ Ministers eventually realized this was far too conservative, following the 2010 introduction of feed-in tariffs, and in 2013 a specific solar strategy [185] was published for the first time, upping the 2020 forecast for PV alone to 10–20 GW.

I am not seeking to lambast my home country, which is certainly not the only one to vacillate about solar policy. Several reports in the Netherlands in the 1980s expressed varying degrees of pessimism, until the 1990 white paper on energy conservation [186] signaled a turnaround in ministerial thinking, saying "PV can become the most important renewable option after 2010."

As mentioned in Section 8.4, there is an argument to suggest that, at least from an economic perspective, there are benefits to late entry.

13.5 Japanese Sunshine Project Rooftop Installations

Table 13.8 summarizes the rooftop installations completed during our time frame by the Sunshine Project.

Table 13.8 Rooftop systems installed by year under Sunshine Project.

Fiscal year	Budget (¥ billion)	Systems no.	Capacity (kW)	Subsidy rate (%)	Ave. Price (¥ million/kW)
1994	2.0	539	1,900	50	2.00
1995	3.3	1,065	3,900	50	1.43
1996	4.1	1,986	7,500	50	1.17
1997	11.1	5,654	19,486	≤33.3	1.03
1998	14.1	6,352	24,123	≤33.3	1.02
1999	16.4	15,879	57,693	≤33.3	0.96
Total	51.0	31,475	114,602	Average	1.03

---

4 Already by 2016, solar projects were supplying power at 12 p/kWh = < 7p in 1991 prices.
5 By the end of 2016, the capacity of PV alone was about 9 GW.

14

When: Conferences, Get-togethers, Prizes

The international PV conferences outlined in Section 5.6 were invaluable rendezvous for the sector, where researchers and industrialists could get together, compare notes, catch up on developments, and, of course, brag a bit.

As time went by, most conference series also spawned one or more awards to give formal recognition to the high achievers, or best self-promoters, of the sector.

This chapter lists the primary international conference series relevant to the terrestrial photovoltaics practitioners, the events occurring during our time frame, and the related prizes and awards.

14.1 IEEE Photovoltaics Specialist Conferences (PVSC)

If proof were needed that this book's start date of 1973 wasn't really the beginning of the global PV industry, the *ninth* IEEE-PVSEC had taken place in Silver Springs, Maryland in 1972, including papers [187] on crystalline silicon by Joseph Lindmayer[qv] and on various cadmium sulfide cells by both Karl Böer[qv] and Dieter Bonnet[qv].

Organized by the Institute of Electrical and Electronic Engineers, this conference series focuses on the scientific and engineering aspects of PV devices and systems for both space and terrestrial applications.

The conferences during our time frame were as follows:

10. 1973 Nov. 13–15; Palo Alto, California
Proceedings [188] 411 pages
Chair: Richard Statler

11. 1975 May 6–8; Scottsdale Arizona
Proceedings [189] 515 pages
Chair: Dennis Curtin

The Solar Generation: Childhood and Adolescence of Terrestrial Photovoltaics, First Edition. Philip R. Wolfe.
© 2018 by the Institute of Electrical and Electronic Engineers, Inc. Published 2018 by John Wiley & Sons, Inc.

12. 1976 Nov. 15–18; Baton Rouge, Louisiana
Proceedings [190]
Chair: Americo Forestieri

13. 1978 June 5–8; Washington DC
Proceedings [191] 1348 pages
Chair: John Goldsmith[qv]

14. 1980 Jan. 7–10; San Diego, California
Proceedings [192]
Chair: Charles Backus
William Cherry Award: Paul Rappaport[qv]

15. 1981 May 12–15; Kissimmee, Florida
Proceedings [193]
Chair: Charles Bishop
William Cherry Award: Joseph Loferski

16. 1982 Sep. 24–30; San Diego, California
Proceedings [194] 1484 pages
Chair: Henry Brandhorst
William Cherry Award: Martin Wolf

17. 1984 May 1–4; Kissimmee, Florida
Proceedings [195] 1432 pages
Chair: Eugene Ralph
William Cherry Award: Henry Brandhorst

18. 1985 Oct. 21–25; Las Vegas, Nevada
Proceedings [196] 1761 pages
Chair: Allen Barnett[qv]
William Cherry Award: Eugene Ralph

19. 1987 May 4–8; New Orleans, Louisiana
Proceedings [197] 1530 pages
Chair: Lawrence Kazmerski[qv]
William Cherry Award: Charles Backus

20. 1988 Sep. 26–30; Las Vegas, Nevada
Proceedings [198]
Chair: Joseph Wise
William Cherry Award: Dave Carlson[qv]

21. 1990 May 21–25; Kissimmee, Florida
Proceedings [199]
Chair: John Meakin
William Cherry Award: Martin Green[qv]

22. *1991 Oct. 7–10; Las Vegas, Nevada*
Proceedings [200]
Chair: Cosmo Barona
William Cherry Award: Peter Iles

23. *1993 May 11–14; Louisville, Kentucky*
Proceedings [201] 1490 pages
Chair: Richard Schwartz
William Cherry Award: Lawrence Kazmerski[qv]

24. *1994 Dec. 5–9; Waikoloa, Hawaii*
This was the 1st World Conference (see page 296)

25. *1996 13–17 May; Piscataway, New Jersey*
Proceedings [202]
Chair: Eldon Boes
William Cherry Award: Allen Barnett[qv]

26. *1997 29 Sep. – 3 Oct.; Anaheim, California*
Proceedings [203] 1451 pages
Chair: Paul Basore
William Cherry Award: Adolf Goetzberger[qv]

27. *1998 Jul. 6–10; Vienna; Austria*
This was the 2nd World Conference (see page 297)
William Cherry Award: Richard Schwartz

14.2 European PV Solar Energy Conferences (EU-PVSEC)

Before the official series of European conferences started, there was a 1974 conference on September 25–27 in Hamburg, which some think of as European Conference Zero [68]. It was primarily a meeting about applications of PV in space, but the last of the eight sessions was on terrestrial applications.

The official European Photovoltaic Solar Energy Conference and Exhibition series was initiated by the European Commission in 1977. WIP was appointed to organize the conferences from 1986 onward. The EU-PVSEC took place every 18 months alternating with the IEEE conference in America, and later also with the Asian conferences.

These conferences started as meetings primarily for researchers in the sector, but evolved in due course to be relevant also to industry, including exhibitions. The format settled down quite early to offer plenary opening and closing sessions, with the rest of the verbal sessions broken into typically about four multiple streams on discrete topic areas. Additionally, further papers were presented in poster sessions. The full proceedings were published in ever larger and heavier tomes; useful not just for reference, but as doorstops too.

The Becquerel Prize was first awarded at the conference in 1989, and Eurosolar periodically awarded its Mouchot Prize here from 1991. Prize winners are shown in the conference listings below.

The conferences held before 2000 were as follows:

1) **1977 Sep. 27–30; Luxembourg**
   Delegates 468 from 31 countries; Proceedings [204] 1378 pages
   Chair: Karlheinz Krebs
   This conference focused on bringing terrestrial PV into the commercial mainstream, for example, with an ambition to reduce costs from "their present price of about \$2000 per square metre" to \$300 and eventually \$50/m^2.
   Promising developments were polycrystalline silicon at about 10% efficiency and thin film cadmium sulfide at 8.6% – compared to monocrystalline then about 13%. Amorphous silicon research was "interesting" as were concentrator systems.

2) **1979 Apr. 23–26; Berlin, West Germany**
   Delegates 520; Proceedings [205] 1210 pages
   Chair: Roger van Overstraeten[qv]
   This conference had increased participation from industry, reflecting growth in the commercial market. In a rueful closing comment, Roger van Overstraeten says "Every conference chairman hopes to see an important technological breakthrough announced at his conference. It did not happen . . ." [205].

3) **1980 Oct. 27–31; Cannes, Franc**
   Delegates 750; Exhibitors 31; Proceedings [206] 1176 pages
   Chair: Wolfgang Palz[qv]
   In a change of running order, the opening sessions covered applications and markets, reflecting how more emphasis had been placed on "system design, current and future applications, goals, markets, and obstacles than at previous conferences" [206]. US production was predicted to rise from 4 MWp in 1980 to 7 MWp in 1981. [207]
   This was the conference where the memorable auction took place to provide solar panels for Mali Aqua Viva (see page 88).

4) **1982 May 10–14; Stresa, Italy**
   Delegates 600; Exhibitors 45; Proceedings [208] 1102 pages
   Chair: Werner Bloss[qv]
   Five days of sunshine greeted this conference, which left an impression of "continued steady progress" [208]. Thin films expected to be the best route to low-cost amorphous silicon and perhaps the most promising, following a record 8% efficient cell reported by Sanyo. Arco Solar announced its plans to build a 1 MW concentrator plant.

5) *1983 Oct. 17–21; Kavouri, Athens, Greece*
Proceedings [209] 1188 pages
Chair: Fabio Fittipaldi
Started with a memorable opening session in the sunshine at an ancient Greek amphitheatre (see Fig. 5.6), addressed by Culture Minister (and former actress) Melina Mercouri.
One of the associated site visits was to the pilot project on Kythnos island, where participants were almost stranded [77].

6) *1985 Apr. 15–19; London, United Kingdom*
Delegates 500; Proceedings [210] 1104 pages
Chair: Fred Treble^{qv}
Started with announcements about the formation of the European Photovoltaic Industry Association^{qv} (EPIA) and the launch of the European Commission's third Non-nuclear Energy Research and Development Programme. The annual world market for PV systems was reported [211] to have reached 25 MW.

7) *1986 Oct. 27–31; Seville, Spain*
Delegates 520 from 39 countries; Proceedings [212] 1266 pages
Chair: Adolf Goetzberger^{qv}
This meeting gave priority to pathways for cost reduction and to the opportunities for PV in developing countries, to which a preconference seminar was devoted. The flamenco at the conference dinner was also memorable, not to mention the bullfight.

8) *1988 May 9–13; Florence, Italy*
Delegates 753 from 43 countries; Exhibitors 37; Proceedings [213] 1725 pages
Chair: Ionel Solomon
The 50% growth in numbers attending, and the sentiment of the participants, reflected a resurgent spirit in the PV sector (reflected by a memorable party at the Intersolar Group^{qv} villa on a vineyard outside Florence). The industry's ability to reach, and beat, the "magic" cost reduction target of $1/W_p$ was reaffirmed. Grid-connected applications were projected to be economic at annual production volumes on the order of 100 MW (global output was then 28 MW). This was the first conference with a session on advanced storage.

9) *1989 Sep. 25–29; Freiburg, Germany*
Proceedings [214] 1217 pages
Chair: Gerard Wrixon
The panel discussion at this conference considered for the first time the social impacts of photovoltaics. This conference was the first at which the Becquerel Prize was awarded.

Becquerel Prize: Roger van Overstraeten[qv]

10) *1991 Apr. 8–12; Lisbon, Portugal*
Delegates 760 from 53 countries; Proceedings [215] 1435 pages
Chair: Antonio Luque[qv]
This was the conference at which Eurosolar's Mouchot Prize was awarded for the first time.
Becquerel Prize: Werner Bloss[qv]
Mouchot Prize: Wolfgang Palz[qv]

11) *1992 Oct. 12–16; Montreux, Switzerland*
Delegates 1,000 from 52 countries; Proceedings [216] 1728 pages
Chair: Leopoldo Guimarães
Becquerel Prize: Antonio Luque[qv]
Mouchot Prize: Charles Imbrecht[qv]

12) *1994 Apr. 11–15; Amsterdam, The Netherlands*
Delegates 1,200 from 65 countries; Proceedings [217] 2192 pages
Chair: Prof. Bob Hill[qv]
With sessions on PV in buildings and countries, this conference again raised the profile of applications for PV alongside cell technology, where several new efficiency records were announced. The opening morning heralded the arrival of a number of solar and electric cars, which had driven from Bonn. Alongside the conference were a two-day symposium on PV for Developing Countries and an Open Day for Architects.
Becquerel Prize: Morton Prince[qv]

13) *1995; Nice, France*
Delegates 1,200; Proceedings [218] 2516 pages
Chair: Werner Freiesleben
Becquerel Prize: Karlheinz Krebs

14) *1997 Jun. 30–Jul 4; Barcelona, Spain*
Delegates from 60 countries; Exhibitors 100; Proceedings [219] 2792 pages
Chair: Heinz Ossenbrink[qv]
Becquerel Prize: Prof. Adolf Goetzberger
Mouchot Prize: John Bonda and Joachim Benemann

15) *1998 Jul. 6–10; Vienna; Austria*
This was the 2nd World Conference (see page 297)
Becquerel Prize: Walter Sandtner

The series of European PV Solar Energy Conferences continues and has latterly become an annual event.
The next conference provided an opportunity for some harmless fun. Paul Maycock published this spoof newsletter on April 1, 2000, telling about the

Figure 14.1 Spoof PV News release [68].

EUROPEAN CONFERENCE CANCELLED

ith warrants issued for the arrest of Hermann Scheer, Bernard McNelis, Heinz C
Palz, Peter Helm and Tapio Alvesalo, Europe's solar community is in crisis wi
ing the 16ᵗʰ EPVSEC, cancelled exactly one month before its scheduled c
Arrests are imminent, and there will undoubtedly be more fall out as the scale c
unfolds. There are rumours that Members of the European Parliament are also i
he first murmurings were thought to be containable, having commenced wher
indicted for serial bullshitology in his claims that nuclear power is bad for your h
al Atomic Energy Agency (IAEA), Vienna, in presenting their case to the Euroj
n Strasbourg, claimed that Dr Scheer had repeatedly made unsubstantiated, ins
nd obnoxious statements that there is recognised danger from electricity gener
ctors. IAEA spokesperson, Prof Dr Frau Rotten Röntgen, in an interview three
ws, said that as the matter was subjudicee she could not comment, but referred e

arrest of Hermann Scheer and others, and the cancellation of the conference (Fig. 14.1).

Some Chinese colleagues, who had spent months getting approval to attend, were unaware of "April Fool's Day" pranks and feared they'd have to cancel their trip.

14.3 Asia-Pacific PV Science and Engineering Conferences

A series of national conferences was held in Japan, starting in 1979, to be superseded by the PVSEC international conference series, which began in 1984.

These conferences were held approximately every 18 months, alternately in Japan and other countries within the Asia/Pacific region. Those held within our time frame were as follows:

1) *1984 Nov. 13–16; Kobe, Japan*
   General Chair: Yoshihiro Hamakawa[qv]
   Program Chair: Kiyoshi Takahashi
   Proceedings [220] 870 pages

2) *1986 Aug. 19–22; Beijing, China*
   General Chair: Zou Xun
   Program Chair: Xu Wen-Yuan
   Proceedings [221] 688 pages

3) *1987 Nov. 3–6; Tokyo, Japan*
   General Chair: Kiyoshi Takahashi
   Program Chair: Taneo Nishino
   Proceedings [222] 816 pages

4) *1989 Feb. 14–17; Sydney Australia*
   General Chair: Martin Green[qv]
   Proceedings [223] 2 volumes

5) **1990 Nov. 26–30; Kyoto, Japan**
   General Chair: Hiroyuki Matsunami
   Program Chair: Masayoshi Umeno
   Proceedings [224] 1050 pages

6) **1992 Feb. 10–14; New Delhi, India**
   General Chair: A. K. Barua
   Program Chair: A. K. Sreedhar
   Proceedings [225] 1125 pages

7) **1993 Nov. 22–26; Nagoya, Japan**
   General Chair: Masayoshi Umeno
   Program Chair: Makoto Konagai
   Proceedings [226] 2 volumes

8) **1995 Feb.; Hawaii, USA**
   This was the 1st World Conference (see page 296)

9) **1996 Nov. 11–15; Miyazaki, Japan**
   General Chair: Makoto Konagai
   Program Chair: Tadashi Saito
   Proceedings [227] 897 pages

10) **1998 Jul. 6–10; Vienna, Austria**
    This was the 2nd World Conference (see page 297)

11) **1999 Sep. 20–24; Sapporo, Japan**
    General Chair: Tadashi Saito
    Program Chair: Takashi Fuyuki
    Proceedings [228] 2 volumes, 1311 pages

This series continues, becoming annual from 2009, and has added South Korea, Thailand, Taiwan, and Singapore as additional countries to host the conference.

14.4 World Conferences on PV Energy Conversion (WCPEC)

The three regional conferences above agreed to combine every 4 years for a World Conference at a location rotating between the three regions, starting in 1994:

1) **1994 Dec. 5–9; Waikoloa, Hawaii**
   This was simultaneously
   24th US IEEE Photovoltaics Specialists Conference
   8th Asia/Pacific PV Science and Engineering Conference
   It was not, on this occasion, considered as an official part of the European series of conferences.

Delegates from 43 countries; Exhibitors 60; Proceedings [229] 2402 pages
Chair: Dennis Flood
Vice Chairs: Masafumi Yamaguchi[qv] and Jürgen Schmid
In addition to the usual technology sessions, this conference covered topics such as financing, installation, operation, and maintenance. Connected events included a public exposition for local residents on the preceding Sunday and a schools program involving 150 students from Hawaii.
Cherry award: Prof. Yoshihiro Hamakawa[qv]

2) *1998 Jul. 6–10; Vienna; Austria*
This was simultaneously
27th US IEEE Photovoltaics Specialists Conference
15th European PV Solar Energy Conference
10th Asia/Pacific PV Science and Engineering Conference
Delegates 2400 from 75 countries; Proceedings [230] 3800 pages
Chairs: Sheila Bailey, Jürgen Schmid, and Kosuke Kurokawa
This reflected the growing aspiration at the end of the last millennium for PV to "become a global player in the energy sector soon" [230]. Several of the keynote presentations dealt with aspirations for 2010, specifying 3000 MW for the EU (half domestically and half in developing countries), 5000 MW for Japan, and $60 billion in the United States.

This series has continued with conferences in Osaka, Hawaii, Valencia, and Kyoto.

14.5 Other Conferences

While the above are the primary conference series specifically devoted to photovoltaics, now mainly for terrestrial applications, many other conferences also cover PV.

14.5.1 International Solar Energy Society

ISES held its first Solar Congress in 1960, and the series has been global and biennial since 1971. The Solar World Congresses held during our time frame were as follows:

1) *1973 Jul. 2–6; Paris, France*
Theme: The Sun in the Service of Mankind

2) *1975 Jul. 28–Aug. 1; Los Angeles, California, USA*
Theme: Solar Use Now – A Resource for People

3) *1978 Jan, 16–21; New Delhi, India*
   Theme: Mankind's Future Source of Energy
   Delegates 800

4) *1979 May 28–Jun 1; Atlanta, Georgia, USA*
   Theme: Silver Jubilee Congress
   Delegates 1100

5) *1981 Aug. 23–28; Brighton, UK*
   Theme: Solar Technology in the Eighties
   Delegates 1200

6) *1983 Aug. 14–19; Perth, Australia*
   Theme: Solar World Congress
   Delegates 1200

7) *1985 Jun. 23–29; Montreal, Canada*
   Theme: Intersol'85
   Delegates 1200

8) *1987 Sep. 13–18; Hamburg, Germany*
   Theme: Advances in Solar Energy Technology
   Delegates 1700 from 81 countries; Proceedings [231] 4 volumes

9) *1989 Sep. 4–8; Kobe, Japan*
   Theme: Clean and Safe Energy Forever
   Delegates 700

10) *1991 Aug. 19–23; Denver, Colorado, USA*
    Theme: Solar Energy for the 21st Century
    Delegates 1400

11) *1993 Aug. 23–27; Budapest, Hungary*
    Theme: Harmony with Nature
    Delegates 1200

12) *1995 Sep. 11–15; Harare, Zimbabwe*
    Theme: In Search of the Sun
    Delegates 565

13) *1997 Aug. 24–30; Taejon, South Korea*
    Theme: Solar Means Business
    Delegates 650

14) *1999 Jul. 4–9 Jerusalem, Israel*
    Theme: Solar is Renewable

14.5.2   World Renewable Energy Congress

Ali Sayigh's World Renewable Energy Network held the first of its biennial Congresses, aimed in particular at broadening knowledge of renewable energy in developing countries, in 1990.

The first three congresses were held in Reading, UK, with the fourth in Denver, USA and the fifth in Florence, Italy in September 1998. During this time the number of delegates increased from 650 to 765. The series continues.

14.5.3   Other PV-related Conferences

Many solar and PV associations – including those mentioned in Section 11.5 – hold their own regular conferences.

Additionally, there is a huge multitude of more focused PV conferences, looking at specific applications, technologies, and geographic regions. There is also a host of renewables conferences, which include sessions on photovoltaics within a broader technology perspective.

14.6   PV Awards and Prizes

The sector has produced a number of awards to recognize achievements in solar photovoltaics, and these are listed further.

A select few innovators have also been recognized beyond the confines of the solar power sector. The only PV-related Nobel Prize to have been awarded, since Einstein's award for achievements in photoelectricity, was to Sir Nevill Mott in 1977 for his work on the electronic structure of magnetic and disordered systems, especially amorphous semiconductors. Some of the pioneers identified herein have also attained election to learned societies and/or received national honors for their achievements.

The PV-specific awards, in the order they were first awarded, are:

14.6.1   IEEE William Cherry Award

The William R. Cherry Award was instituted in 1980, shortly after his death, to recognize an engineer or scientist who devoted a part of their professional life to the advancement of the science and technology of photovoltaic solar energy conversion. The nominee must have made significant contributions to the science and/or technology of PV energy conversion, with dissemination by substantial publications and presentations.

It is presented by the IEEE at its PV Specialists Conference. Pre-2000 winners were as follows:

1980 Paul Rappaport[qv]

1981 Joseph Loferski

1982 Martin Wolf

1984 Henry Brandhorst

1985 Eugene Ralph

1987 Charles Backus

1988 David Carlson[qv]

1990 Martin Green[qv]

1991 Peter Iles

1993 Lawrence Kazmerski[qv]

1994 Yoshihiro Hamakawa[qv]

1996 Allen Barnett[qv]

1997 Adolf Goetzberger[qv]

1998 Richard Schwartz

Subsequent winners include Chris Wronski, Dick Swanson[qv], Antonio Luque[qv], Masafumi Yamaguchi[qv], and NREL's Tim Coutts, Jerry Olson, and Keith Emery.

14.6.2 Becquerel Prize

Edmond Becquerel was, let's face it, a rather obscure French scientist, although he is revered by solar aficionados as the first to observe the photovoltaic effect. It was a surprise, therefore, to find the BBC making a documentary on the 150th anniversary of this discovery to coincide with the London conference[qv].

Bob Hill[qv] is credited [43] with the suggestion of using this opportunity to inaugurate the Alexandre Edmond Becquerel Prize to honor long-term scientific, technical, or managerial achievement in the development of photovoltaic solar energy.

It is primarily a European Award but not restricted exclusively to European citizens. The prizes are awarded at the European PV Specialist Conferences, and the pre-2000 winners were as follows:

1989 Prof. Roger van Overstraeten[qv]

1991 Prof. Werner Bloss[qv]

1992 Prof. Antonio Luque[qv]

1994 Dr. Morton Prince[qv]

1995 Dr. Karlheinz Krebs[qv]

1997 Prof. Adolf Goetzberger^{qv}

1998 Dr. Walter Sandtner

Subsequent winners include Fred Treble^{qv}, Wolfgang Palz^{qv}, Masafumi Yamaguchi^{qv}, Dieter Bonnet^{qv}, Arvind Shah, and Dick Swanson^{qv}.

14.6.3 PVSEC Awards

The PVSEC Award is awarded at each Asia/Pacific PVSEC Conference^{qv} to a person from the PVSEC society who has devoted many years to the progress of photovoltaic science and engineering. Winners pre-2000 were as follows:

1996 Yoshihiro Hamakawa^{qv}

1998 Yukinori Kuwano^{qv}

1999 Makoto Konagai

Subsequent winners include Masafumi Yamaguchi^{qv}.
PVSEC also periodically makes Special Awards for people who have made a major contribution to the growth of photovoltaic industrial technology or public policy. In our time frame, these went to the following:

1996 Tadashi Sasaki

1999 Kazuo Inamori

Subsequent winners include Shi Zhengrong and Osamu Ikki.
This conference series has latterly added the Hamakawa Award for scientific achievement.

14.6.4 Mouchot Prize

The Augustin Mouchot Prize is named after the nineteenth century French inventor of an early solar-powered steam engine. It was created by Eurosolar to be "awarded annually to individuals or organizations that have created the basis for a solar civilization in their respective fields of science, industry, communication, and politics." It is often awarded during the European PV Conferences, and the pre-2000 winners were as follows:

1989 John Bockris and Eduard Justi

1991 Dr. Wolfgang Palz^{qv}

1992 Dr. Charles Imbrecht

1994 Bruno Topel^{qv}

1997 John Bonda^{qv} and Joachim Benemann^{qv}

14.6.5 Karl W Böer Solar Energy Medal of Merit

The prize is awarded biennially for making significant pioneering contributions to solar energy conversion. The award comes with a substantial financial prize from the University of Delaware-managed foundation named in honor of Karl Böer^{qv}. Winners from its inception to the end of the century were as follows:

1993 President Jimmy Carter

1995 David Carlson^{qv}

1997 Adolf Goetzberger^{qv}

1999 Stanford Ovshinsky^{qv}

Subsequent winners include Allen Barnett^{qv}, Martin Green^{qv}, Yoshihiro Hamakawa^{qv}, Larry Kazmerski^{qv}, Hermann Scheer^{qv}, Dick Swanson^{qv}, and Antonio Luque^{qv}.

14.6.6 Bonda Prize

Created by the European Photovoltaic Industry Association^{qv} in memory of its late Director General, the Bonda Prize was first awarded at the Glasgow Conference^{qv} in 2000 to Claude Remy^{qv}.

Subsequent winners include the architect of Berlin station (and its PV glass roof), Christoph-Friedrich Lange; Peter Varadi^{qv}; Hans-Josef Fell^{qv}, and Winfried Hoffman^{qv} in 2012, after which it seems to have lapsed.

14.6.7 Other Awards

ISES, the International Solar Energy Society, and its national chapters give awards individually within the broader solar sector, as mentioned in its profile in Section 11.5. These have gone to PV people including Werner Bloss^{qv}, Adolf Goetzberger^{qv}, and subsequently Bernard McNelis^{qv} and Larry Kazmerski^{qv}.

Similarly, the World Renewable Energy Network gives awards in the renewables sector, and these have sometimes recognized PV pioneers such as Masafumi Yamaguchi^{qv}.

Latterly, the Robert Hill Award for the Promotion of PV for Development has been created in memory of Bob Hill, and recipients include Anil Cabraal, Bernard McNelis^{qv}, and Bunker Roy.

NREL^{qv} initiated the Paul Rappaport Renewable Energy and Energy Efficiency Award in 2002 in memory of Paul Rappaport^{qv}.

14.6.8 A Footnote About Prizes

Awards can be a mixed blessing in any sector, and PV is no exception. There is a danger that the awards committee becomes unduly cliquey or "politically correct," while the need to choose just one recipient every year can lead to deserving individuals missing out in some years and less prominent contributors being feted in others.

Indeed most recipients of the above awards are giants of the sector, as can be seen in Chapter 10. But, for example, none of the US famous five early entrepreneurs Bill Yerkes, Peter Varadi, Dr. Ishaq Shahryar, Joseph Lindmayer, and Elliot Berman received a major award, apart from Peter's Bonda Prize. Nor did any woman receive a major PV award during our time frame, and only Sarah Kurtz and Mechtild Rothe have been recognized subsequently.

Part III

Dictionary, References, Glossary, and Indexes

> *"This revolution, the information revolution, is a revolution of free energy as well, but of another kind: free intellectual energy."*
>
> Steve Jobs [232]

This third part of the book is designed for those who want to exercise their "free intellectual energy."

It doesn't add to the story of how the PV sector developed during its first quarter century. But it does show where the information needed to write that story came from. And it leads those, who want to know more, to a treasure trove of further interesting resources.

Don't sit down to read it from beginning to end!

The Solar Generation: Childhood and Adolescence of Terrestrial Photovoltaics, First Edition. Philip R. Wolfe.
© 2018 by the Institute of Electrical and Electronic Engineers, Inc. Published 2018 by John Wiley & Sons, Inc.

A

Acknowledgments and Reminiscences

I am hugely grateful to the collective wisdom of many who have helped me along the way. I hope they will all forgive me if I omit herein all their well-earned doctorates, professorships, engineering qualifications, and so on.

Interviews, Discussions, and Correspondence

First, my thanks go to those who kindly gave up their time to tell me about their experiences, and in particular:

A1	Bill and Sara Yerkes [154]; Santa Barbara, CA, USA; October 2011
A2	Allen Barnett; University of New South Wales, Australia; November 2014
A3	Martin Green [124]; University of New South Wales, Australia; November 2014
A4	Bernard and Zhu Li McNelis [68]; Devon, UK; April and October 2016
A5	James Watson; Solar Power Europe, Brussels; April 2016
A6	Joachim Benemann [69]; Cologne, Germany; April 2016
A7	Winfried Hoffmann [99]; Henau, Germany; April 2016
A8	Michael Schmela; Munich, Germany; April 2016
A9	Peter Helm; Munich, Germany; April 2016
A10	Heinz Ossenbrink; JRC Ispra, Italy; April 2016
A11	Claude and Olena Remy [142]; Lyon, France; April 2016
A12	Robert and Françoise de Franclieu [122]; Lyon, France; April 2016
A13	Michael Grätzel [123]; Lausanne, Switzerland; April 2016
A14	Wolfgang Palz; Paris, France; May 2016
A15	Bruce Cross; Pontypridd, UK; June 2016
A16	Emmanuel Fabre; from Avignon, France; July 2016
A17	Roberto Vigotti; from Rome, Italy: July 2016

A18	Giovanni Simoni [163]; from Rome, Italy; August 2016
A19	Jeremy Leggett; from Kent, UK; August 2016
A20	Dipesh Shah [93]; London, UK; August 2016
A21	Tapio Alvesalo; Helsinki, Finland; August 2016
A22	Adolf Goetzberger [170]; Freiburg, Germany; August 2016
A23	Murray Cameron; from Munich, Germany; August 2016
A24	Nicky Pearsall [126]; Newcastle, UK; August 2016
A25	Elliot Berman [117]; New York, USA; September 2016
A26	Earle Kazis; New York, USA; September 2016
A27	Karl and Renate Böer [119]; Naples, FL, USA; September 2016
A28	Neville and Patti Williams [35]; Naples, FL, USA; September 2016
A29	Paul and Roma Maycock [86]; Williamsburg, VA, USA; September 2016
A30	Dave and MaryAnn Carlson [63]; Williamsburg, VA, USA; September 2016
A31	Peter and Helene Varadi [150]; Chevy Chase, USA; September 2016
A32	Scott Sklar [101]; Washington, DC, USA; September 2016
A33	Keith Emery [171]; NREL, Golden, CO, USA; September 2016
A34	Charlie Gay [49]; Los Angeles, CA, USA; September 2016
A35	Raju Yenamandra [29]; Thousand Oaks, CA, USA; September 2016
A36	Sara Yerkes; Santa Barbara, CA, USA; September 2016
A37	Peter Aschenbrenner [24]; Sunpower, San Jose, CA, USA; September 2016
A38	Dick Swanson [69]; Stanford, CA, USA; September 2016
A39	John Day [42]; Los Altos, CA, USA; September 2016
A40	Paula Mints [76]; San Jose, CA, USA; September 2016
A41	Bob and Marie Johnson [34]; Rio Vista, CA, USA; September 2016
A42	Jim and Jan Caldwell [75]; Sonoma, CA, USA; September 2016
A43	Guy and Simone Smekens [149]; Brussels, Belgium; September 2016
A44	Godfrey Bevan; London, UK; October 2016
A45	Anthony Derrick [46]; Devon, UK; November 2016
A46	Hubert Aulich; from Erfurt, Germany; March 2017
A47	Heather Harper; Iver, UK; May 2017

Hopefully, they will forgive me for including partners' names – especially as in many cases, they added highly perceptive views of the high and low lights of the early PV years to complement those of the interviewee!

I was fortunate also to have access to the video interviews [103] by 10 of the participants to the Cherry-Hill Conference (see Section 8.3).

Private Communications

Thanks are also due to others, who have provided their views, reminiscences, or feedback through electronic communications, including

A48	Theresa Jester; December 2016
A49	Roger Little; January 2017
A50	Yukinori Kuwano [130]; January 2017
A51	Larry Kazmerski (Kaz); February 2017
A52	John Goldsmith; February 2017
A53	Markus Real [40]; June 2017
A54	Gene Ralph; December 2017

Quotations and insights specific to individual respondents are referenced in the text and linked to the citations in the following chapter.

However, most of the content of the foregoing book is a distillation from reminiscences of all those mentioned above – combined with my own memories – much of which is not specifically cited below.

Image Credits

Many of the people and organizations mentioned have been kind enough to provide photographs. I am grateful to Neville Williams [35], Markus Real [40], Paul Maycock [86], Yukinori Kuwano [130], Anthony Derrick [46], John Day [42], Joachim Benemann [69], and to Giovanni Simoni for providing early EPIA publications [163].

Many of the photographs of organizations in Chapter 11 and individuals in Section 10.2 came from those institutions and people themselves [25]. I am also grateful for others whose images are included: Associated Press [47], Miguel Brullet [39], Energy Efficiency [28], German Information Centre [160], Los Alamos National Laboratory [71], National Renewable Energy Laboratory [176], Polenet [21], Sue Roaf [36], Royal Signals and Radar Establishment [169], Shell International [79], Wiki-Solar [45], and Wikimedia Commons [26].

Most especially, a large number of images and papers have come from Bernard McNelis's extensive archive [68] and from Intersolar Group [27]. Uncaptioned images came from one of these sources, or the organization or individual shown [25].

B

Cited References

The following publications are referenced in the text. International Standard Book Numbers (ISBN) use the 10-digit format (for the 13-digit format, add 978- at the start).

A bibliography that extracts some of the most suitable references for further reading is given in Chapter C below.

1 Rollo Appleyard, Letter to the editor. Telegraphic Journal and Electrical Review, 1981.

Referenced in Volume I

2 Editorial in New York Times following demonstration of "the first successful device to convert useful amounts of the sun's energy directly and efficiently into electricity" by Bell Telephone Laboratories at the National Academy of Sciences in Washington DC. April 26, 1954.

3 Charles Lutwidge Dodgson (under the pseudonym Lewis Carroll), *Alice in Wonderland*. London: Macmillan, 1865.

4 Albert Einstein, Über einen die Erzeugung und Verwandlung des Lichtes betreffenden heuristischen Gesichtspunkt [On a heuristic viewpoint concerning the production and transformation of light]. *Annalen der Physik*, **17**, 132–148, 1905.

5 Edmond Becquerel, Mémoire sur les effets électriques produits sous l'influence des rayons solaires. *Comptes Rendus*, **9**, 561–567, 1839. Available at http://gallica.bnf.fr/ark:/12148/bpt6k2968p/f561 (accessed Aug. 2016). (Impressively, thanks to the Bibliothèque Nationale de France.)

6 US Department of Energy, The History of Solar, 2002. Available at https://www1.eere.energy.gov/solar/pdfs/solar_timeline.pdf (accessed October 2016).

7 M. King Hubbert, Nuclear Energy and the Fossil Fuels, Presented before the Spring Meeting of the Southern District, American Petroleum Institute, Plaza

The Solar Generation: Childhood and Adolescence of Terrestrial Photovoltaics, First Edition. Philip R. Wolfe.
© 2018 by the Institute of Electrical and Electronic Engineers, Inc. Published 2018 by John Wiley & Sons, Inc.

Hotel, San Antonio, Texas, Shell Oil Company/American Petroleum Institute, March 7–9, 1956.

8 Noel Grove, Emory Kristof, Oil: the Dwindling Treasure, National Geographic (June).

9 Kjell Aleklett, *Peeking at Peak Oil*. New York: Springer, 2012, ISBN 1461434245.

10 Keith de Lacy, Solar and Wind Power Simply Don't Work – Not Here, Not Anywhere, opinion piece, The Australian, June 22, 2016.

11 Alfred Smee, *Elements of Electro-Biology, or the Voltaic Mechanism of Man of Electro-Pathology, Especially of the Nervous System and of Electro-Therapeutics*. London: Longman, Brown, Green & Longmans, 1849, p. 15. ("Upon exposing the apparatus to intense light, the galvanometer was instantly deflected, shewing that the light had set in motion a voltaic current, which I propose to call a photo-voltaic circuit.")

12 Heinrich Hertz, Uber den Einfluss des ultravioletten Lichtes auf die electrische Entladung. *Annalen der Physik*, **267** (8) 983–1000, 1887.

13 John Perlin, *From Space to Earth: the Story of Solar Electricity*. Boston: Harvard University Press, 1999, ISBN 0937948144.

14 Solar Cell History. Available at http://an-surzglobal.com/html/solar_cell_history.html (accessed November 2016).

15 Vladimir Zworykin, Edward Ramberg, *Photoelectricity and its applications*. New York: John Wiley & Sons, Inc., 1949, ISBN 1559180560.

16 John Perlin, The invention of the solar cell. Popular Science, April 22, 2014.

17 Daryl Chapin, Calvin Fuller, Gerald Pearson, Solar energy converting device, U. S. Patent US2780765, filed March 5, 1954 and published February 5, 1957.

18 John Perlin, The story of Vanguard. Extract from From space to earth, the story of solar electricity [13]. Available at http://beta.deepspace.ucsb.edu/outreach/the-space-race/the-story-of-vanguard (accessed October 2016).

19 Karl Böer, Future large scale terrestrial use of solar energy. Proceedings of 25th Power Sources Symposium, Atlantic City, N.J., May 23–25, 1972, PSC Publications Committee, Red Bank.

20 Surface Area Required to Solar Power the World. 2009. Available at http://www.informationisbeautiful.net/2009/surface-area-required-to-solar-power-the-world/ (accessed October 2016).

21 Polenet, Image credit, 2008.

22 Soteris Kalogirou (ed.) *McEvoy's Handbook of Photovoltaics: Fundamentals and Applications: Third Edition*. Oxford: Elsevier, 2017.

23 Jan Czochralski, Ein neues Verfahren zur Messung der Kristallisationsgeschwindigkeit der Metalle [A new method for the measurement of the crystallization rate of metals]. *Zeitschrift für Physikalische Chemie*, **92**, 219–221, 1918.

24 Peter Aschenbrenner, Discussion (see A37 above), 2016.

25 Organisations and individuals named (1975–2017). Images provided.

26 Wikimedia Commons, Library of publicly available images, 2017.

27 Intersolar Group^{qv} and Solapak, (1981–2001) Image library available to the author.

28 Energy Efficiency, Image credit, 1998.

29 Raju Yenamandra, Discussion (see A35 above), 2016.

30 Jacquelyn Ottman, We are all green consumers, now and for the future. Environmental Leader, February 17, 2011.

31 *Surface Area Required to Solar Power the World*. Available at http://www. informationisbeautiful.net/2009/surface-area-required-to-solar-power-the-world/ (accessed October 2016), 2009. (The authors added this footnote, "According to the UN 170,000 square kilometres of forest is destroyed each year. If we constructed solar farms at the same rate, we would be finished in 3 years.")

32 Gordon Raisbeck, The solar battery. *Scientific American*, 1955, **193**/6 (Dec).

33 Graham Teale, Philip Wolfe, System sizing: the theory and the practice, Proceedings of Fifth E.C. Photovoltaic Solar Energy Conference C (See also Ref. [209]), 417–423, 1983.

34 Bob Johnson, Discussion (see A41 above), 2016.

35 Neville Williams, Images and discussion (2016 see A28 above), 1980s.

36 Sue Roaf, Image credit, 1993.

37 Adolf Goetzberger, Volker Hoffmann, *Photovoltaic Solar Energy Generation*. Heidelberg: Springer, 2005.

38 100 000 Roofs Solar Power Programme, Global Renewable Energy Policies and Measures Database, International Energy Agency and International Renewable Energy Agency. Available at http://www.iea.org/textbase/pm/? mode=re&id=28&action=detail (accessed February 2012).

39 Miguel Brullet, Image credit, 1999.

40 Markus Real, Image and communications (2017 see A53 above), 1981.

41 Nelson Mandela, Address at the launch of the rural non-grid electrification by the joint venture between Shell Solar and Eskom, Bhipa, Flagstaff District February 24, 1999.

42 John Day, Images and discussion (2016 see A39 above), 1983.

43 Wolfgang Palz, *Solar Power for the World: What You Wanted to Know About Photovoltaics*. Singapore: Pan Stanford Publishing, 2014, ISBN 9814411875.

44 Philip Wolfe, *Solar Photovoltaic Projects in the Mainstream Power Market*. Oxford: Routledge, 2013, ISBN 0415520485.

45 Wiki-Solar Database and image library available to author and online (2012–2017). Available at http://wiki-solar.org/index.html (accessed April 2017).

46 Anthony Derrick, Image and correspondence (see A45 above), 2016.

47 Associated Press, Image credit, 1960.

48 Trent Lott, quoted in the Washington Post, May 23, 1997.

49 Charles Gay, Discussion (see A34 above), 2016.

50 John Brockman, A Theory of Roughness: A talk with Benoit Mandelbrot. 2004. Available at https://www.edge.org/conversation/benoit_mandelbrot-a-theory-of-roughness (accessed March 2017). (Author's note: As an avid devotee of Chaos Theory in general, and – see Fig. B.1 – the Mandelbrot set in particular, I am delighted to have found this almost relevant quote from the great man.)

Fig. B.1 Mandelbrot set extract author's office [27].

51 Peter F. Varadi, *Sun Above the Horizon, Meteoric Rise of the Solar Industry.* Singapore: Pan Stanford Publishing, 2014, ISBN 9814613293.
52 University of New South Wales, Cross section of buried contact solar cell, 1991.
53 Michael Lammert, Richard Schwartz, The interdigitated back contact solar cell: a silicon solar cell for use in concentrated sunlight. *IEEE Transactions on Electron Devices*, **24** (4), 337–342, 1977.
54 Andres Cuevas, The early history of bifacial solar cells, Proceedings of the 20th EPVSEC, Barcelona, 2005.
55 Richard Bube, *Photovoltaic Materials.* London: Imperial College Press, 1998, ISBN 1860940651.
56 Dietrich Jenny, Joseph Loferski, Paul Rappaport, Photovoltaic effect in GaAs p–n junctions and solar energy conversion. *Physical Review*, **101**, 1208–1209, 1956.
57 Martin Green, *Third Generation Photovoltaics: Advanced Solar Energy Conversion.* Heidelberg: Springer, 2003, ISBN 3540265634.
58 D.C. Reynolds, G. Leies, L.L. Antes, Photovoltaic effect in cadmium-sulphide. *Physical Review* **96**, 533, 1954.
59 Jet Propulsion Laboratory, *Photovoltaic Conversion of Solar Energy for Terrestrial Applications* (October 23–25, 1973 Cherry Hill, New Jersey).

California: Jet Propulsion Laboratory, California Institute of Technology 1974.

60 Gerhard Willeke, Eicke Weber, *Advances in Photovoltaics*. Oxford: Elsevier, 2012, ISBN 0123884190.

61 Dominic Cusano, CdTe solar cells and photovoltaic heterojunctions in II–VI compounds. *Solid State Electronics*, **6**, 217–232, 1963.

62 David Carlson, Semiconductor device having a body of amorphous silicon. U. S. Patent US4064521A, filed July 30, 1976 and published December 30 1977.

63 David Carlson, Discussion (see A30 above), 2016.

64 Yukinori Kuwano et al. 8% efficiency a-SiC:H/a-Si:H heterojunction solar cells. Proceedings of the Fourth European Photovoltaic Conference 698–703, 1982.

65 Lawrence Kazmerski, F.R. White, G.K. Morgan, Thin film CuInSe2/CdS heterojunction solar cells. *Applied Physics Letters*, **29**, 268, 1976.

66 Helmut Tributsch, Reaction of excited chlorophyll molecules at electrodes and in photosynthesis. *Photochemistry and Photobiology*, **16**, 4, 1972.

67 Brian O'Regan, Michael Grätzel, A low-cost, high efficiency solar cell based on dye-sensitized colloidal TiO2 films. *Nature*, **353**, 737–740, 1991.

68 Bernard McNelis, (1975–1999) Images, discussion (2016 see A4 above).

69 Joachim Benemann, Images and discussion (2016 see A6 above), 1990.

70 Dick Swanson, Discussion (see A38 above), 2016.

71 Los Alamos National Laboratory Image credit, 2014.

72 George Löf, John Duffie, Clayton Smith, *World Distribution of Solar Radiation*. Madison: University of Wisconsin, 1966.

73 Philip Wolfe, System sizing-further advances towards the definitive solution, *Sixth E.C. Photovoltaic Solar Energy Conference: Proceedings of the International Conference, held in London, UK, 15–19 April 1985*, Dordrecht: D. Reidel 64–70, 1984, ISBN 90-277-2104-4.

74 Amory Lovins, Kyle Datta, *Winning the Oil Endgame: Innovation for Profits, Jobs and Security*. Boulder: Rocky Mountain Institute, 2005, ISBN 1881071105.

75 James Caldwell, Discussion (see A42 above), 2016.

76 Paula Mints, Discussion (see A40 above), 2016.

77 John Day, An Excursion to Kythnos – October 22 1983, Memoir by John Day, Los Altos. 2016 Available at http://solargeneration.pub/library/ Day_Kythnos.pdf (accessed May 2017).

78 Hal Macomber, Application Trends for Photovoltaics, *Fourth E.C. Photovoltaic Solar Energy Conference, Proceedings of the International Conference, held at Stresa, Italy, 10–14 May 1982*. Dordrecht: D. Reidel, 30–39, 1982, ISBN 94-009-7900-0.

79 Shell International (early 1990s); no longer available from Shell, but included in: Sue Roaf, Manuel Fuentes, and Stephanie Thomas, *Ecohouse: A Design Guide*. Oxford: Architectural Press, 2001, ISBN 0750649046.

80 Michael Starr, Small-scale, stand-alone photovoltaic systems – an overview. Proceedings of the Symposium on Materials and New Processing Technologies for Photovoltaics, Electrochemical Society, 1985.

81 Energy Technology Support Unit, *Solar Energy: Its Potential Contribution within the United Kingdom, Energy Paper Number 16*, London: Department of Energy/HMSO, 1977, ISBN 0114102944.

82 Bob Johnstone, *Switching to solar*. New York: Prometheus, 2010, ISBN 1616142223.

83 Wikipedia, Moore's Law. Available at https://en.wikipedia.org/wiki/Moore%27s_law (accessed March 2017).

84 Richard Swanson, A vision for crystalline silicon photovoltaics. *Progress in Photovoltaics, Research and Applications*, **14**, 443–453, 2006.

85 Wikipedia, Swanson's Law. Available at https://en.wikipedia.org/wiki/Swanson%27s_law (accessed March 2017).

86 Paul Maycock, Papers and discussion (2016 see A29 above), 1980.

87 Tim Bruton et al. Multi-Megawatt Upscaling of Silicon and Thin Film Solar Cell and Module Manufacturing 'MUSIC FM' (EU contract APAS RENA CT94 0008), BP Solar, Sunbury, 1996.

88 John Perlin, Solar power: the slow revolution. *Invention and Technology*, **18**, 1, 2001.

89 Leonard Magid, Current status of the U.S. terrestrial photovoltaic conversion programme, *Photovoltaic Solar Energy Conference, Proceedings of the International Conference held at Luxembourg September 27–30 1977*. Dordrecht: D. Reidel, 1977, 453–461, ISBN 9027708892.

90 US Department of Energy, *National Photovoltaics Program, Electrical power from solar cells*. Pasadena: Jet Propulsion Laboratory, September, 1980.

91 Robert Hershey, Solar power race at Jersey concern, New York Times. July 2, 1982.

92 President George W Bush, comment reportedly made to UK Prime Minister Tony Blair after a discussion with French Prime Minister Georges Chirac at the G8 Summit, 2002.

93 Dipesh Shah, Discussion (see A20 above), 2016.

94 John Browne, Robin Nuttall, Tommy Stadlen, *Connect: How Companies Succeed by Engaging Radically with Society*. London: W H Allen, 2016, ISBN 0753556948.

95 Arnaud Chaperon,"Le cœur de métier de Total est la production d'hydrocarbures mais cela ne nous empêche pas de mener une réflexion sur les énergies au sens large du terme", interview *Plein Soleil*, January 20, 2012.

96 Exxon Corporation Advertisement,"Exxon answers questions about one of the newest sources of energy under the sun – the sun", November, 1976.

97 John Keyes, *The Solar Conspiracy*. New York: Morgan & Morgan, 1975 ISBN 0871000958.

98 Theresa Jester, From Pet Project to Partner, O&G Investment in Solar. Renewable Energy World, August 23, 2016.

99 Brian Farmer, Howard Wenger, Thomas Hoff, Charles Whitaker, Performance and value analysis of the Kerman 500 kW photovoltaic power plant, Proceedings of the American Power Conference. April 1995.

100 Winfried Hoffmann, Discussion (see A7 above), 2016.

101 Scott Sklar, Discussion (see A32 above), 2016.

102 Sir George Porter, Sayings of the Week, The Observer. London, August 26, 1973.

103 Video interviews, Cherry Hill Revisited – An Historical Remembrance: The Beginning of the U.S. PV Revolution. 1999.

104 US Congress, Public Utility Regulatory Policies Act of 1978, Public Law 95-617, Washington DC, November 9, 1978.

105 Glenn Strahs, Carol Tombari, *Laying the Foundation for a Solar America: The Million Solar Roofs* Initiative. Washington DC: US Department of Energy, 2006.

106 Dennis Costello, *Photovoltaic Procurement Strategies: An Assessment of Supply Issues*. Golden: Solar Energy Research Institute, 1980.

107 Michael Starr, Karlheinz Krebs, *Photovoltaic Activities at the Joint Research Centre, Ispra: A Review of Results, Achievements and Plans*. Brussels: European Commission, 1988.

108 Based on IEA data from Energy Technology RD&D Report (2016 edition) © OECD/IEA (2015), www.iea.org/statistics. Licence, www.iea.org/t&c, as analysed by the author.

109 Markus Real, Pierre de Ruvo, Richard Kay, Peter F. Varadi, Photovoltaics, quality assured. Renewable Energy World, 2004.

110 Mahatma Gandhi, *Freedom's Battle, Being a Comprehensive Collection of Writings and Speeches on the Present Situation*. Madras: Ganesh & Co, 1922.

111 United Nations Environment Programme Green Economy and Trade – Trends, Challenges and Opportunities. 2013.

112 Paul Maycock, Vic Sherlekar, *Photovoltaic Technology, Performance Cost and Market Forecast to 1995*. Alexandria, VA: Photovoltaic Energy Systems, 1984.

113 Paul Maycock, International photovoltaic markets, developments and trends forecast to 2010, *Proceedings of First World Conference on Photovoltaic Energy Conversion, Twenty Fourth IEEE Photovoltaic Specialists Conference 1994. Waikoloa, HI, USA, 5–9 December 1994*. New York: IEEE, 1994, ISBN 0-7803-1460-3.

114 Kurzweil, Ray Steven Bushong, Futurist Ray Kurzweil predicts solar industry dominance in 12 years, Solar Power World. 2016. Available at http://www .solarpowerworldonline.com/2016/03/futurist-ray-kurzweil-predicts-solar-industry-dominance-12-years/ (accessed March 2017).

Referenced in Volume II

115 James Newton, *Uncommon Friends: Life with Thomas Edison, Henry Ford, Harvey Firestone, Alexis Carrel, and Charles Lindbergh.* Dublin: Houghton Mifflin Harcourt, 1989, ISBN 0156926201.

Who's Who

116 George Bernard Shaw, *Man & Superman, Maxims for Revolutionists.* Cambridge: The University Press, 1903, ISBN 1587340569.
117 Elliot Berman, Discussion (see A25 above), 2016.
118 Werner Bloss, *Elektronische Energiewandler.* Stuttgart: Wissenschaftliche Verlagsges, 1968 (ASIN B0000BQ5CE).
119 Karl Böer, Discussion (see A27 above), 2016.
120 Karl Wolfgang Böer, *Handbook of the Physics of Thin-Film Solar Cells.* Berlin: Springer Verlag, 2013, ISBN 3642367472.
121 Karl Wolfgang Böer with Esther Riehl, *The Life of the Solar Pioneer Karl Wolfgang Böer.* New York: iUniverse, 2010, ISBN 1450228787.
122 Robert de Franclieu, Discussion (see A12 above), 2016.
123 Michael Grätzel, Discussion (see A13 above), 2016.
124 Martin Green, (2014–2016) Discussion (see A3 above) and other communications.
125 Mary Archer, Martin Green, *Clean Electricity from Photovoltaics.* Singapore: World Scientific, 2014, ISBN 1783266708.
126 Nicola Pearsall, Discussion (see A24 above), 2016.
127 Robert Hill, *Prospects for Photovoltaics, Commercialization, Mass Production and Application for Development.* New York: United Nations, 1992, ISBN 9211043913
128 Winfried Hoffmann, *The Economic Competitiveness of Renewable Energy: Pathways to 100% Global Coverage.* Chichester: John Wiley & Sons, Ltd, 2014, ISBN 1118237908.
129 Barbara Fox, Michele Alperin, Solar power for Have-Nots, US1 Newspaper, September 6, 2006.
130 Yukinori Kuwano, Images and correspondence (2017 see A50 above), 1976, 1993.

131 Walter Spear, Peter LeComber, Electronic properties of substitutionally doped amorphous Si and Ge. Philosophical Magazine, 1976, London **33**/6 (received for publication February 20).

132 Joseph Lindmayer, Chuck Wrigley, *Fundamentals of Semiconductor Devices.* D Van Nostrand: Princeton, 1966, ISBN 0442048076.

133 Antonio Luque, Steven Hegedus, *Handbook of Photovoltaic Science and Engineering.* Chichester: John Wiley & Sons, Ltd, 2003.

134 A. Luque, A. Martí, Increasing the efficiency of ideal solar cells by photon induced transitions at intermediate levels. *Physical Review Letters*, **78** (26) 5014–5017, 1997.

135 Paul Maycock and Edward Stirewalt, *Photovoltaics: Sunlight to Electricity in One Step.* Massachusetts: Brick House Publishing, 1981.

136 Jonathan Fahey, Repeat pretender. Forbes Magazine, November 24, 2003.

137 Joann Muller, Stanford Ovshinsky, battery genius behind smartphones and hybrids, dies at 89. Forbes Magazine, October 18, 2012.

138 Wolfgang Palz, *Solar Electricity: An Economic Approach to Solar Energy.* Oxford: Butterworth, 1978, ISBN 0408709101.

139 Morton Prince, Silicon solar energy converters. *Journal of Applied Physics*, **26**, 534, 1955.

140 US Signal Corps, Newsletter of the Association for Applied Solar Energy. 1960.

141 Markus Real, *Wie kamm die Sonne ins Netz? [How the sun got into the grid].* Wisconsin: Books on Demand, 2017, ISBN 3743159945

142 Claude Remy, Discussion (see A11 above), 2016.

143 Claude Remy, Photowatt – Solar France, l'historique de Photowatt, le plus factuel possible, Private communication, Lyon, August 2015. Available at http://solargeneration.pub/library/Remy_Photowatt.pdf (accessed May 2017).

144 Alain Ricaud, Les politiques publiques de l'énergie solaire 1973–2013, Cythelia, Le Bourget du Lac. Available at http://www.cythelia.fr/images/Image/file/Politiques%20publiques%20PV%201973-2013%20Alain%20Ricaud%20v3.pdf (accessed May 2017).

145 Hermann Scheer, *Der energethische Imperativ, 100% jetzt, Wie der vollständige Wechsel zu erneuerbaren Energien zu realisieren ist.* Munich: Verlag Antje Kunstmann, 2010, ISBN 3888976834.

146 Hermann Scheer, *Sonnen-Strategie – Politik ohne Alternative.* Munich: Piper Verlag, 1999 (ASIN B00CJUJG72).

147 Hermann Scheer, *Solare Weltwirtschaft, Strategie für die ökologische Moderne [The Solar Economy].* Munich: Verlag Antje Kunstmann, 2005, ISBN 3888973147

148 Fred Pearce, Interview, The sun king. New Scientist, April 13, 2002.

149 Guy Smekens, Discussion (see A43 above), 2016.

150 Peter Varadi, Discussion (see A31 above), 2016.

151 Peter F. Varadi, *Sun Towards High Noon*. Singapore: Pan Stanford Publishing, 2014.

152 Neville Williams, *Chasing the Sun: Solar Adventures around the world*. Vancouver: New Society Publishers, 2005, ISBN 1550923124.

153 Neville Williams, *Sun Power: How Energy from the Sun is Changing Lives around the World, Empowering America and Saving the Planet*. New York: Doherty Associates, (2014)

154 Bill Yerkes, Discussion (see A1 above), 2011.

155 Bill Yerkes: the Henry Ford of Photovoltaics. SolarWorld. 2015. Video available at https://www.youtube.com/watch?v=mKuJayD6z3A (accessed February 2018).

156 John J Berger, *Charging Ahead: The Business of Renewable Energy and What it Means for America*. Berkeley: University of California Press, 1998, ISBN 0520216143.

157 Isaac Newton, Letter to Robert Hooke, February 15, 1676.

158 Shiela Bailey, Larry Viterna, and K.R. Rao, *Role of NASA in Photovoltaic and Wind Energy*. New York: American Society of Mechanical Engineers, 2011.

159 Don Felder, Don Henley, and Glenn Frey, Hotel California, Asylum Records, 1977.

Companies and Organizations

160 German Information Centre, Image credit, 1982.

161 Business News, Arco wins fraud suit stemming from 1990 unit sale to Siemens. Wall Street Journal, December 9, 1997.

162 Edgar Gunther, AstroPower Corporation, Decline of a solar photovoltaic star remains an untold tragedy? *Gunther Portfolio*. 2006. Available at http://guntherportfolio.com/2006/08/astropower-decline-of-a-solar-photovoltaic-star-remains-an-untold-tragedy/ (accessed October 2016).

163 Giovanni Simoni, European Photovoltaic Industry Association (first yearbook) provided in discussion (2016 see A18 above), 1985.

164 Kyocera website www.kyocerasolar.com, 2016.

165 Sharp website www.sharp-world.com/solar, 2016.

166 Gordon McKeague, The Washington Post, August 9, 1983.

167 Eugene Ralph, Photovoltaic technology assessment, Proceedings of 18th IEEE Photovoltaics Specialists Conference, 1985. (Also see Ref. [196].)

168 John Day III, Status and future aspects of photovoltaics, *Advances in Solar Energy Technology: Proceedings of the Biennial Congress of the International Solar Energy Society Hamburg, Federal Republic of Germany, 13–18 September 1987*. vol. 1, Oxford: Pergamon Press, 1985, 27–33.

169 Royal Signals and Radar Establishment, Image credit, 1978.

170 Adolf Goetzberger, Discussion (see A22 above), 2016.

171 Keith Emery, Discussion (see A33 above), 2016.

172 SERI, *A New Prosperity, Building a Sustainable Energy Future: The Seri Solar Conservation Study*. Golden: Solar Energy Research Institute, 1981.

173 Martin Green, et al. Improvements in silicon solar cell efficiency, *Conference Record of the 18th IEEE Photovoltaic Specialists Conference, Las Vegas, Nevada, October 21–25*. New York: IEEE, 1985, 39–42.

174 Anil Cabraal, Mac Cosgrove-Davies, Loretta Schaeffer, Best Practices for Photovoltaic Household Electrification Programs, Lessons from Experiences in Selected Countries, World Bank Technical Paper 324, Washington: World Bank, 1996.

175 Anil Cabraal, Mac Cosgrove-Davies, Loretta Schaeffer, Accelerating sustainable photovoltaic market development. *Progress in Photovoltaics*, **6**, 297–306, 1998.

Technology, Geography, and Politics

176 National Renewable Energy Laboratory, Best Research Cell Efficiencies (see its website – C36 below) http://nrel.gov/pv/assets/images/efficiency_chart .jpg, 2017.

177 William Shockley, Hans Queisser, Detailed balance limit of efficiency of p–n junction solar cells. *Journal of Applied Physics*, **32**, 510, 1961.

178 Henry Brandhorst, *Photovoltaics: The Endless Spring, NASA Technical Memorandum 83684*. Cleveland: Lewis Research Center, 1984.

179 European Commission, *Photovoltaic Power Generation, Proceedings of the Final Design Review Meeting on EC Photovoltaic Pilot Projects held in Brussels, Nov. 30–Dec. 2 1981*, Dordrecht: D. Reidel, 1982.

180 Bill Gillett, Rod Hacker, Willi Kaut, *Photovoltaic Demonstration Projects, Proceedings of the fifth contractors' meeting*. Brussels: European Commission, 1991.

181 UK, Department of Energy, Energy Research and Development in the United Kingdom, Energy Paper Number 11, London: HMSO, 1976.

182 Department of Trade & Industry, New and Renewable Energy, Future Prospects in the UK, Energy Paper Number 62, London: HMSO, 1994.

183 John Battle (Energy Minister), House of Commons Written Answers for 31 Jul 1998, Hansard, London, 1998.

184 Department of Energy and Climate Change, *The UK Renewable Energy Strategy*. London: HMSO, 2009.

185 Department of Energy and Climate Change, *UK Solar PV Strategy Part 1: Roadmap to a Brighter Future*. London: HMSO, 2013.

186 *Nota Energiebesparing* (1990) (White Paper on Energy Saving). Dutch Ministry of Economic Affairs, The Hague.

Proceedings – IEEE Photovoltaic Specialist Conferences

187 IEEE, Proceedings of the 9th IEEE Photovoltaic Specialists Conference, Silver Spring 1972, New York: IEEE, 1973.

188 IEEE, Conference Record of the 10th IEEE Photovoltaic Specialists Conference, Nov. 13–15 1973, New York: IEEE, 1974.

189 IEEE Conference Record of the 11th IEEE Photovoltaic Specialists Conference, May 6–8 1975, Scottsdale, Arizona, IEEE, New York, 1975.

190 IEEE Conference Record of the 12th IEEE Photovoltaic Specialists Conference, Nov 15–18 1976, Baton Rouge Louisiana, New York: IEEE, 1977.

191 IEEE Conference Record of the Thirteenth IEEE Photovoltaic Specialists Conference, June 5–8 1978, Washington, New York: IEEE, 1978.

192 IEEE Conference Record of the 14th IEEE Photovoltaic Specialists Conference, January 7–10 1980, San Diego, California, New York: IEEE, 1980.

193 IEEE Conference Record of the 15th IEEE Photovoltaic Specialists Conference, Kissimmee, Florida, May 12–15 1981, New York: IEEE, 1981.

194 IEEE Conference Record of the 16th IEEE Photovoltaic Specialists Conference, Sept. 27–30 1982, San Diego, Calif., New York: IEEE, 1982.

195 IEEE Conference Record of the seventeenth IEEE Photovoltaic Specialists Conference 1984, Hyatt-Orlando Hotel, Kissimmee, Florida, May 1–4 1984, New York: IEEE, 1984.

196 IEEE Conference Record of the 18th IEEE Photovoltaic Specialists Conference, Las Vegas, Nevada, October 21–25, New York: IEEE, 1985.

197 IEEE Conference Record of the nineteenth IEEE Photovoltaic Specialists Conference 1987, Sheraton Hotel, New Orleans, Louisiana, May 4–8 1987, New York: IEEE, 1987.

198 IEEE Conference Record of the 20th IEEE Photovoltaic Specialists Conference, NV, USA, 26–30 Sep 1988, New York: IEEE, 1988.

199 IEEE Proceedings of Twenty First IEEE Photovoltaic Specialists Conference 1990, Kissimmee, FL, USA, 21–25 May 1990, New York: IEEE, 1990.

200 IEEE Proceedings of Twenty Second IEEE Photovoltaic Specialists Conference 1991, Las Vegas, NV, USA, 7–11 Oct. 1991. New York: IEEE, 1992, ISBN 0-87942-636-5.

201 IEEE The Conference Record of the Twenty Third IEEE Photovoltaic Specialists Conference 1993, Louisville, KY, USA, 10–14 May 1993, New York: IEEE, 1993, ISBN 0-7803-1220-1.

202 IEEE Conference Record of the Twenty-Fifth IEEE Photovoltaic Specialists Conference 1996, New York: IEEE, 1996.

203 IEEE Conference Record of the Twenty Sixth IEEE Photovoltaic Specialists Conference 1997, Anaheim Marriott, Anaheim, CA, 29 September–03 October 1997, New York: IEEE, 1997.

Proceedings – European PV Solar Energy Conferences

204 Albert Strub (ed) Photovoltaic Solar Energy Conference, Proceedings of the International Conference held at Luxembourg September 27–30 1977, Dordrecht: D. Reidel, 1978, ISBN 9027708892.

205 Roger van Overstraeten, Wolfgang Palz, Gerhard Willeke (eds) 2nd E.C. Photovoltaic Solar Energy Conference, Proceedings of the International Conference, held at Berlin (West), 23–26 April 1979, Dordrecht: D. Reidel, 1979, ISBN 90-277-1021-5.

206 Wolfgang Palz (ed) Third E.C. Photovoltaic Solar Energy Conference, Proceedings of the International Conference, held at Cannes, France, 27–31 October 1980, Dordrecht: D. Reidel, 1981, ISBN 9400984257.

207 Paul Maycock, Photovoltaics program overview, *Third E.C. Photovoltaic Solar Energy Conference, Proceedings of the International Conference, held at Cannes, France, 27–31 October 1980*, Dordrecht: D. Reidel, 10–17, 1981.

208 Prof. Werner Bloss, Dr. Guiliano Grassi (eds) Fourth E.C. Photovoltaic Solar Energy Conference, Proceedings of the International Conference, held at Stresa, Italy, 10–14 May 1982, Dordrecht: D. Reidel, 1982, ISBN 94-009-7900-0.

209 Wolfgang Palz, Fabio Fittipaldi (eds), Fifth E.C. Photovoltaic Solar Energy Conference, Proceedings of the international conference held at Kavouri (Athens), Greece, 17–21 October 1983, Dordrecht: D. Reidel, 1984, ISBN 90-277-1724-5.

210 Frederick Treble, Wolfgang Palz (eds) Sixth E.C. Photovoltaic Solar Energy Conference, Proceedings of the international conference, held in London, UK, 15–19 April 1985, Dordrecht: D. Reidel, 1985, ISBN 90-277-2104-4.

211 Paul Maycock, The current PV scene worldwide. *Sixth E.C. Photovoltaic Solar Energy Conference, Proceedings of the international conference, held in London, UK, 15–19 April 1985*, Dordrecht: D. Reidel, 21–27, 1985.

212 Adolf Goetzberger, Wolfgang Palz, Gerhard Willeke, (eds) Seventh E.C. Photovoltaic Solar Energy Conference, Proceedings of the International Conference, held at Sevilla, Spain, 27–31 October 1986, Dordrecht: D. Reidel, 1987, ISBN 90-277-1724-5.

213 Ionel Solomon, Bernard Equer, Peter Helm, (eds) Eighth E.C. Photovoltaic Solar Energy Conference, Proceedings of the International Conference, held at Florence, Italy, 9–13 May 1988, Dordrecht: D. Reidel, 1988, ISBN 90-277-2815-9.

214 Gerard Wrixon, Wolfgang Palz, Peter Helm (eds) Ninth E.C. Photovoltaic Solar Energy Conference, Proceedings of the international conference, held at Freiburg, 25–29 September 1989, Dordrecht: Kluwer Academic, 1989, ISBN 0-7923-0497-7.

215 Antonio Luque, Gabriel Sala, Wolfgang Palz, Gonçalves dos Santos (eds) Tenth E.C. Photovoltaic Solar Energy Conference, Proceedings of the International Conference, held at Lisbon, Portugal, 8–12 April 1991, Dordrecht: Springer, 1991, ISBN 94-010-5607-6.

216 Leopoldo Guimarães (ed) Eleventh E.C. Photovoltaic Solar Energy Conference, Proceedings of the International Conference, held at Montreux, Switzerland, 12–16 October 1992, Harwood: Reading Academic, 1993, ISBN 37-186-5380-5.

217 Robert Hill, Wolfgang Palz, Peter Helm (eds) Twelfth European Photovoltaic Solar Energy Conference, Proceedings of the International Conference, held at Amsterdam, the Netherlands 11–15 April 1994. London: H.S. Stephens, 1994, ISBN 09-521-452-40.

218 Werner Freiesleben, Peter Helm, Heinz Ossenbrink, Wolfgang Palz (eds) Thirteenth European Photovoltaic Solar Energy Conference, Proceedings of the International Conference, held at Amsterdam, the Netherlands 11–15 April 1994, London: James & James, 1996, ISBN 09-521-4527-1.

219 Heinz Ossenbrink, Peter Helm, Heinz Ehmann (eds) Fourteenth European Photovoltaic Solar Energy Conference, Proceedings of the International conference held at Barcelona, Spain, 30 June–4 July 1997, London: H.S. Stephens, 1998, ISBN 19-016-7503-0.

Proceedings – Asia/Pacific Conferences

220 Yoshihiro Hamakawa (ed), Technical Digest 1st International Photovoltaic Science and Engineering Conference, Kobe, Japan, Nov. 13–16 1984, Tokyo (NCID BA37888381), 1984.

221 Yu Pei-Nuo (ed), Proceedings of the 2nd International Photovoltaic Science and Engineering Conference, Aug. 19–22, 1986, Beijing, China, Adfield Advertising, Hong Kong (NCID BA70051393), 1986.

222 Kiyoshi Takahashi (ed), Technical Digest, 3rd International Photovoltaic Science and Engineering Conference, Nov. 3–6/1987, Tokyo, Japan, Keidanren Kaikan, Secretariat of International PVSEC-3, Tokyo (NCID BA50421106), 1987.

223 PVSEC-4 Proceedings of the 4th International Photovoltaic Science and Engineering Conference, February 14–17 1989, Sydney, NSW Australia, Sydney: The Institution of Radio and Electronic Engineers Australia, 1989, ISBN 0909394164.

224 International PVSEC-5 Technical Digest, 5th International Photovoltaic Science and Engineering Conference, Nov. 26–30/1990, Kyoto, Japan, Miyako Hotel, Kyoto: International PVSEC-5, 1990 (NCID BA70052250).

225 B. K. Das, S. N. Singh (eds) 6th International Photovoltaic Science and Engineering Conference, Proceedings, February 10–14, 1992, New Delhi, India, New Delhi: Oxford & IBH Pub. Co., 1992, ISBN 8120406796.

226 Yoshihiro Hamakawa (ed) Proceedings of the 7th International Photovoltaic Science and Engineering Conference, Nagoya, Japan, 22–26 November 1993, New York: Elsevier, 1994.

227 PVSEC-9 Technical Digest, 9th International Photovoltaic Science and Engineering Conference, Nov. 11–15/1996, Miyazaki, Japan, Seagaia Convention Complex, World Convention Center Summit, Tokyo: International PVSEC-9, 1996 (NCID BA89407216).

228 T. Fuyuki, T. Saitoh (eds) Special issue: PVSEC 11, Amsterdam: Elsevier, 2001.

Proceedings – World Conferences on PV Energy Conversion

229 IEEE Proceedings of First World Conference on Photovoltaic Energy Conversion, Twenty Fourth IEEE Photovoltaic Specialists Conference 1994. Waikoloa, HI, USA, 5–9 December 1994, New York: IEEE, 1995, ISBN 0-7803-1460-3.

230 Jürgen Schmid et al. (eds) 2nd World Conference on Photovoltaic Solar Energy Conversion, Proceedings of the International Conference held at Vienna, Austria on 6–10 July 1998 (Vol I of 3), Ispra: European Commission, 1998, ISBN 92-828-5179-6.

Proceedings – Other Conferences

231 Werner Bloss, Fritz Pfisterer (eds) Advances in Solar Energy Technology: Proceedings of the Biennial Congress of the International Solar Energy Society Hamburg, Federal Republic of Germany, 13–18 September 1987, Oxford: Pergamon Press, 1988, ISBN 0-08-034315-5.

Referenced in Volume III

232 Steven Jobs, A candid conversation about making computers, making mistakes and making millions, Playboy Magazine, Chicago, February, 1985.

233 Clayton Christensen, *The Innovator's Dilemma: When New Technologies Cause Great Firms to Fail.* Boston: Harvard Business Review, 1997, ISBN 1422154637.

234 Philip Wolfe, The Solar Generation, Available at http://www.solargeneration .pub 2017.

Further Bibliography

235 Herbert Girardet, Miguel Mendonça, *A Renewable World*. Totnes: Green Books, World Future Council, 2009.

236 Rebecca Hill, *Applications of Photovoltaics*. Boca Raton: CRC Press, 1989, ISBN 0852742778.

237 Richard F Hirsch, *Technology and Transformation in the American Electric Utility Industry*. New York: Cambridge University Press, 1989.

238 Jeremy Leggett, The Winning of the Carbon War (self-published), 2016.

239 Amory Lovins, *Reinventing Fire: Bold Business Solutions for the New Energy Era*. Colorado: Rocky Mountain Institute, 2011.

240 Pere Mir-Artigues, Pablo del Río, *The Economics and Policy of Solar Photovoltaic Generation*. Switzerland: Springer, 2016, ISBN 3319296531.

241 John Perlin, *Let It Shine: The 6,000-Year Story of Solar Energy*. San Francisco: New World Library, 2013, ISBN 1608681327.

242 Hans Rauschenbach, *Solar Cell Array Design Handbook: The Principles and Technology of Photovoltaic Energy Conversion*. New York: Van Nostrand Reinhold, 1980, ISBN 9401179171.

243 Chetan Singh Solanki, *Solar Photovoltaics: Fundamentals, Technologies and Applications*. New Delhi: PHI Learning, 2009.

C

Bibliography: Books, Publications, and Websites

There are so many references noted in the text and listed in Chapter B, so I have extracted a shorter list that may be useful for those interested in further reading.

The books listed in the first section below have been very helpful in researching this one. I have referred to them at relevant places in the text, thereby avoiding the temptation to plagiarize.

The Solar Generation: Childhood and Adolescence of Terrestrial Photovoltaics, First Edition. Philip R. Wolfe.
© 2018 by the Institute of Electrical and Electronic Engineers, Inc. Published 2018 by John Wiley & Sons, Inc.

Books About the Photovoltaic Industry and Sector

C1   Bob Johnstone's *Switching to Solar* [82]

C2   Wolfgang Palz's *Solar Power for the World* [43], which I have referred to as 'the *Solar Power* book'

C3   John Perlin's books, notably *From Space to Earth, The Story of Solar Electricity* [13]

C4   Peter Varadi's books [51,151]

C5   Neville Williams' books [152,153]

Selected Photovoltaic Textbooks

C6   Mary Archer and Martin Green's *Clean Electricity from Photovoltaics* [125]

C7   Karl Böer's *Handbook of the Physics of Thin-Film Solar Cells* [120]

C8   Adolf Goetzberger's *Photovoltaic Solar Energy Generation* [37]

C9   Antonio Luque's *Handbook of Photovoltaic Science and Engineering* [133]

C10   On utility-scale PV: *Solar Photovoltaic Projects in the Mainstream Power Market* [44]

See also the further titles listed under "Further Bibliography" on page 326.

Istiography[1]

This is a far from exhaustive list of websites (prefix with http://www. unless otherwise listed) of useful sources and entities mentioned in this book.

Surviving/Successor PV Companies in Sections 11.2 and 11.3

C11   Applied Materials, acquirer of HCT Shaping Systems, appliedmaterials.com

C12   Crystalox, now PV Crystalox, pvcrystalox.com

---

1 My own word: Bibliography comes from the Greek "biblios" for book, "istios" is a web.

C13 ENEL Green Power, the energy company's renewables business, enelgreenpower.com

C14 EniPower, the successor organization to what started as Pragma, enipower.it

C15 ITPEnergised acquirers of IT Power, itpenergised.com

C16 Kyocera, kyocerasolar.com

C17 Panasonic, panasonic.net/ecosolutions/solar

C18 Photowatt now owned by EDF, photowatt.com

C19 Sharp, sharp-world.com/solar

C20 SMA Solartechnik, sma.de

C21 Solar Electric Light Fund, self.org

C22 Solar Frontier, new name for Showa-Shell's business, solar-frontier.com

C23 Solarworld, the organization that acquired what was Arco Solar then Siemens Solar then Shell Solar, solarworld.com

C24 Spectrolab, spectrolab.com/solarcells

C25 SunPower Corporation, sunpower.com

C26 Total, total.com/en/energy-expertise/exploration-production/solar-power

C27 Wacker Chemie, wacker.com

C28 WIP, wip-munich.de

Research Establishments Profiled in Section 11.4

C29 Delaware Institute of Energy Conversion, udel.edu/iec

C30 École Polytechnique Fédérale de Lausanne, epfl.ch

C31 EC Joint Research Centre, ec.europa.eu/jrc

C32 Fraunhofer ISE, ise.fraunhofer.de

C33 IMEC at Katholiek University of Louvain, imec-int.com

C34 Jet Propulsion Laboratory, jpl.nasa.gov

C35 Madrid Technical University Institute of Solar Energy, ies.upm.es

C36 National Renewable Energy Laboratory, nrel.gov

C37 Northumbria University NPAG, northumbria.ac.uk/about-us/academic-departments/mathematics-physics-and-electrical-engineering/research/northumbria-photovoltaic-applications (easy winner of the longest web address!)

C38   Stanford University, stanford.edu

C39   Stuttgart University Institute of Physical Electronics, uni-stuttgart.de/forschung/orp/inst_profile/fak05/ipe.en.html

C40   UNSW-SPREE, engineering.unsw.edu.au/energy-engineering

National and International Bodies Profiled in Section 11.5

C41   Eurec, eurec.be

C42   European Commission, europa.eu

C43   European Photovoltaic Industry Association, now called Solar Power Europe, solarpowereurope.org

C44   IEA – International Energy Agency, iea.org and iea-pvps.org

C45   IEEE, ieee.org

C46   International Solar Energy Society, ises.org

C47   MITI, now METI – Ministry of Economy, Trade and Industry (Japan), meti.go.jp

C48   SEIA – Solar Energy Industries Association (US), seia.org

C49   US DOE, energy.gov/eere/sunshot/sunshot-initiative

C50   World Bank, worldbank.org and scalingsolar.org

Other PV Industry Associations

C51   Assosolare – Italy, assosolare.org

C52   BSW – Bundesverband Solarwirtschaft (German Federal Solar Industry Association), solarwirtschaft.de

C53   CanSIA – Canadian Solar Industries Association, cansia.ca

C54   SER – Syndicat des énergies renouvelables, France, enr.fr

C55   JPEA – Japan Photovoltaic Energy Association, jpea.gr.jp

C56   SESI – Solar Energy Society of India, sesi.in

C57   CPIA – China Photovoltaic Industry Association, chinapv.org.cn

Surviving Conferences and Awards Profiled in Chapter 14

C58   IEEE Photovoltaics Specialist Conferences (IEEE-PVSC), ieee-pvsc.org

C59   European PV Solar Energy Conferences (EU-PVSEC), photovoltaic-conference.com

C60   Asia/Pacific PV Science and Engineering Conferences, pvsec.org

C61   World Renewable Energy Congress, wrenuk.co.uk

C62   IEEE William Cherry Award, ieee-pvsc.org/PVSC43/awards-cherry.php

C63   Becquerel Prize, becquerel-prize.org

C64   PVSEC Awards, pvsec.org/awards.html

C65   Karl W Böer Solar Energy Medal of Merit, udel.edu/iec/karlboeraward

Other Useful Sources

C66   *The Solar Generation*, solargeneration.pub

C67   IREnA – International Renewable Energy Agency, irena.org

C68   RTS Corporation, rts-pv.com

C69   Sandia National Laboratories (US DOE), sandia.gov

C70   Wiki-Solar, resource on utility-scale PV, wiki-solar.org

D

Glossary, Units, Conversions, and Standards

However hard you try, the jargon has to creep in eventually. This section contains the following:

Abbreviations (Including Organization Names)

This does not include technical units – such as kWh and MW_P – which are given under Section "Units, Abbreviations, and Conversion Factors."

AC	Alternating current
AFME	L'Agence Française pour la Maîtrise de l'Energie – the French government's energy agency, latterly replaced by L'Agence de l'Environnement et de la Maîtrise de l'Energie (ADEME)
AIM	The Alternative Investment Market of the London Stock Exchange
AIST	Institute of Advanced Industrial Science and Technology – a national laboratory in Japan
Amoco	Formed when the American Oil Company was acquired by Standard Oil of Indiana
Arco	The Atlantic Richfield Company
ASIF	Asociación de la Industria Fotovoltaica, the Spanish Industry Association

The Solar Generation: Childhood and Adolescence of Terrestrial Photovoltaics, First Edition. Philip R. Wolfe.
© 2018 by the Institute of Electrical and Electronic Engineers, Inc. Published 2018 by John Wiley & Sons, Inc.

ASTAE	The Asia Sustainable and Alternative Energy Program of the World Bank
BBC	British Broadcasting Corporation
BIPV or BiPV	Building-integrated photovoltaics, using solar components specifically designed to form part of the building structure, such as solar roof tiles, façades, and brise soleils
BMFT	Bundesministerium für Forschung und Technologie – the German Ministry for Research and Technology (now part of the Ministry of Education and Research)
BSI	British Standards Institution – the UK's national body for the establishment of standards
CanSIA	Canadian Solar Industries Association
CEC	The Commission of the European Community – the administrative government of the EEC; latterly the EC
CEO	Chief Executive Officer
CFCs	Chlorofluorocarbons
CILSS	Comité Permanent Inter-Etats de lutte contre la Sécheresse dans le Sahel – covers Burkina Faso, Cape Verde, Gambia, Guinea-Bissau, Mali, Mauritania, Niger, Sénégal, and Tchad
COMES	Commissariat à l'Energie Solaire – the French solar energy agency
CNRS	Centre National de la Recherche Scientifique – a French government agency
CSP	Concentrated solar power – uses tracking mirrors or lenses to concentrate sunlight onto a boiler or tube, usually to produce high-temperature steam for power generation
CVD	Chemical vapor deposition
DC	Direct current
DFID	Department for International Development of the UK government
EC	The European Commission – the administrative government of the EU; formerly the CEC
ECD	Energy Conversion Devices
EEC	The European Economic Community – has subsequently become the European Union

EEG	Erneuerbare-Energien-Gesetz (Renewable Energy Act) – the German legislation introducing, *inter alia*, feed-in tariffs for renewable electricity, as further described in Section 13.4
EFG	Edge-defined, film-fed growth – a method of producing ribbon silicon for solar cells, typically in octagonal form
ENE	Energies Nouvelles et Environnement – the Belgian company
ENEA	Energia Nucleare ed Energie Alternative – an Italian sustainable energy agency that is now known as Agenzia Nazionale per le Nuove Tecnologie, l'Energia e lo Sviluppo Economico Sostenibile
ENEL	Ente Nazionale per l'Energia Elettrica – the Italian national electricity company
ENI	Ente Nazionale Idrocarburi – the Italian national oil company
EPIA	The European Photovoltaic Industry Association – now Solar Power Europe
ETH Zurich	The Eidgenössische Technische Hochschule in Zürich
ETSU	The Energy Technology Support Unit – a former UK government agency
EU	The European Union – it has emerged out of the EEC
EVA	Ethylene vinyl acetate
HIT	Heterojunction with intrinsic thin layer
HMSO	Her Majesty's Stationery Office – official government publisher in the United Kingdom
IAEA	The International Atomic Energy Agency
IEA	The International Energy Agency^{qv} – an autonomous intergovernmental organization established in Paris in 1974 by the OECD
IEC	The International Electrotechnical Commission – an NGO responsible for International Standards for electrical, electronic, and related technologies
IECEE	IEC System of Conformity Assessment Schemes for Electrotechnical Equipment and Components
IEEE	The Institute of Electrical and Electronic Engineers^{qv} – in the United States, *inter alia*, organizers of the US PV Specialist Conferences
IMEC	The Interuniversity Micro-Electronica Centrum^{qv} at KUL

ISO	The International Standards Organization – administers the establishment and maintenance of international standards
ITC	The Business Energy Investment Tax Credit – enacted under US tax law, as described in Section 8.3
IUPAC	International Union of Pure and Applied Chemistry – responsible for element nomenclature (e.g., III–V, see Glossary of Terms below)
IWES	Institut für Windenergie und Energiesystemtechnik (the Institute for Wind Energy and Energy Systems Technology) – part of the Fraunhofer Institute^{qv}
JPEA	Japan Photovoltaic Energy Association
KfW	Originally named the Kreditanstalt für Wiederaufbau – this is a German government-owned development bank
KUL	Katholieke Universiteit Leuven^{qv}
LCOE	The levelized cost of energy – the total capital and operating cost of the plant divided by the energy output, as discussed in Section 6.4 and defined in Glossary of Terms below
MBB	Messerschmitt-Bölkow-Blohm – the German-based company
MBE	Member of the Order of the British Empire
MBO	Management buyout – the acquisition of a company by a group of its employees from the previous parent company
MIS	Metal–insulator–semiconductor
MPPT	Maximum power point trackers
NASA	The National Aeronautics and Space Administration in the United States
NASDAQ	National Association of Securities Dealers Automated Quotations – a US stock exchange
NATO	North Atlantic Treaty Organization
NCPV	The US National Center for Photovoltaics, based at NREL^{qv}
NEC	Nippon Electric Corporation
NEDO	New Energy Industrial Technology Development Organisation of Japan
NGO	Nongovernmental organization – a noncommercial or not-for-profit organization independent of states or international governmental organizations

NIST	National Institute of Science and Technology – under the US Department of Commerce
NIST-ATP	The Advanced Technology Program of NIST – mentioned in Section 8.3
NOCT	Normal operating cell temperature for solar module testing – see Section 12.3
NPAC	Newcastle Photovoltaic Applications Centre^{qv} – at the University of Northumbria, UK
NREL	The National Renewable Energy Laboratory^{qv} – under the US Department of Energy, headquartered in Golden, Colorado
NTT	Nippon Telegraph and Telephone Public Corporation
OAPEC	The Organization of Arab Petroleum Exporting Countries
OBE	Officer of the Order of the British Empire
OPEC	The Organization of Petroleum Exporting Countries
PG&E	Pacific Gas and Electric
PIRDES	Programme Interdisciplinaire de Recherche pour le Développement de l'Energie Solaire – launched by CNRS in 1975
PR	Public relations
PTC	The Renewable Electricity Production Tax Credit – enacted under US tax law, as described in Section 8.3
PURPA	The US Public Utility Regulatory Policies Act – first enacted in 1978
PV	Photovoltaics
PVB	Polyvinyl butyral
PVF	Polyvinyl fluoride
PVMaT	The US Department of Energy's PV Materials and Technology Program – managed by NREL, see Section 8.3
PVT	Hybrid photovoltaic thermal devices that convert solar radiation to both electricity and heat (not to be confused with TPV)
qv	*Quod vide* – Latin, literally "which see"; used for entities described elsewhere in this book. In the case of people, in Chapter 10; companies and research organizations in Chapter 11; conferences and awards in Chapter 14
R&D; RD&D	Research and development; research, development, and demonstration

RAE	The Royal Aircraft Establishment in the United Kingdom
REC	Renewable Energy Corporation of Norway – with spin-offs REC Solar and REC Silicon
RPS	Renewable Portfolio Standard – see Glossary of terms below
RV	Recreational vehicle – such as caravan or motor home
SCADA	Supervisory control and data acquisition
SER	Syndicat des Energies Renouvelables – French industry association
SERI	The Solar Energy Research Institute – forerunner of NREL^{qv}
SESI	Solar Energy Society of India
SHW	Solar hot water (not a photovoltaic application)
SPC	Solar Power Corporation^{qv}
SPREE	The School of Photovoltaic & Renewable Energy Engineering – at the University of New South Wales^{qv}
SQ limit	The Shockley–Queisser efficiency limit – see Section 12.2
STC	Standard test conditions – see Section 12.3
Toe	Tonne of oil equivalent
TPV	Thermophotovoltaics – whereby sunlight (or another heat source) is used to heat an intermediate body whose glow is captured by a PV solar cell and converted into electrical power (not to be confused with PVT)
UMIST	University of Manchester Institute of Science and Technology
UN	United Nations
UNCED	UN Conference on Environment and Development
UNDP	UN Development Programme
UNFCCC	UN Framework Convention on Climate Change – the administrative body for the Kyoto Protocol and subsequent climate agreements
UNSW	The University of New South Wales^{qv}

Glossary of Terms

II–VI semiconductors	Made from compounds comprising elements from columns II and VI of the periodic table of elements (IUPAC nomenclature now calls this 12–16)

III–V semiconductors	Compounds comprising elements from columns III and V of the periodic table of elements (now 13–15)
IV	Elements in column IV (now 14) of the periodic table, sometimes called the "carbon group"
air mass	The optical path length for sunlight through the atmosphere, relative to that at sea level with the sun directly overhead
amorphous silicon; a-Si; α-Si	Allotropic form of silicon, usually deposited as a thin film, in such a way as to have no regular or crystalline structure
array; solar array	A collection of mounted and interconnected solar modules on a mounting structure
availability *or* availability factor	The percentage of the time that electricity-generating plant is available to produce electricity (not to be confused with capacity factor)
balance-of-system (BOS)	Parts of a PV generating system apart from the solar modules
capacity	The power rating of a solar cell, module, or system at which it would operate under Standard Test Conditions
capacity factor	The ratio of the electrical energy produced by a generating unit relative to the electrical energy that could have been produced at continuous full-power operation during the same period
climate change	The long-term variations in global and regional climates, now commonly attributed to the impact of mankind
crystalline silicon	Silicon material produced so that it has an ordered atomic structure, either monocrystalline or polycrystalline
degression	Periodic decreases in feed-in tariffs; designed to match falling costs with reductions in the level of support
diffuse radiation	That part of the incident solar radiation that has been diffused by particulate in the atmosphere, such as clouds

direct radiation

Light incident on a surface that comes directly from the sun without having been reflected or diffused

disruptive technology

A technology or product that displaces an established technology and disrupts the existing industry or creates a completely new one; coined by Clayton Christensen [233]

distributed generation

Power generation embedded in (connected to) an electricity distribution network (as compared to centralized generation usually connected to the transmission network)

distribution network

The network for electricity distribution at voltages below that of the transmission network

double-axis tracker

A tracker that follows the sun in its daily orbit and adjusts to the height of the sun at different seasons

early PV era

The period between 1973 and 1999, being the primary focus for this book.
Also referred to as *our time frame* or *the first solar generation*; additionally, the words *latterly* and *subsequently* are frequently applied to developments after the early PV era

Earth Summit

The 1992 UNCED (see Abbreviations above)

efficiency (of solar cells and modules)

The ratio of the electrical output to the light energy input

energy

The cumulative power output of a system over a period of time, measured in watt hours or equivalent units

energy efficiency

The amount of useful output achieved for each unit of input energy. Energy efficiency can be improved by reducing both the wastage within the equipment and its use

energy mix

The distribution of different energy sources within the total supply

energy security

The extent to which a nation's or region's energy supplies are ensured against potential disruption, including factors such as depletion of natural resources, variability, and the political stability of regions where the energy supplies are obtained or transshipped

feed-in tariff (FiT)	A renewable energy support mechanism, where producers are rewarded at a level prescribed by legislation for renewable electricity produced and/or fed into the grid
fill factor	A measure of the shape or "squareness" of an I–V characteristic (see Section 2.3), calculated as

$$\frac{\text{Peak power}(W_P)}{\text{Open-circuit voltage } V_{OC} \times \text{short-circuit current } I_{SC}}$$

	where these parameters are all measured under the same test conditions
first oil crisis	The first OPEC oil embargo, described in Section 1.1
first solar generation	The period between 1973 and 1999, being the primary focus for this book. Also referred to as *our time frame* or *the early PV era*; additionally, the words *latterly* and *subsequently* are frequently applied to developments after our time frame
flash test	A performance test on a solar module by flashing it with light to simulate STC radiation conditions
free-field	A solar system supported on the ground as opposed to building-mounted or building-integrated systems
global radiation	The total solar resource falling on a surface, being the combination of the direct and the diffuse radiation
greenhouse gases (GHG)	Gases whose emission contributes to climate change; the Kyoto Protocol identifies these as carbon dioxide, methane, nitrous oxide, hydrofluorocarbons, perfluorocarbons, and sulfur hexafluoride
grid-connected (of solar power systems)	Configured to deliver electricity to the transmission or distribution network
grid parity	The point at which the cost per kilowatt-hour of solar power matches that of traditional electricity. See also *retail grid parity* and *wholesale grid parity*
ibid	Ibidem (Latin – *in the same place*) – papers and articles included in the composite work last listed above
insolation	The level of light illumination on a surface

istiography (a word invented by the author)	List of websites (istiography is to websites as bibliography is to books)
I–V characteristic	The curve showing the relationship between the current output of a solar cell or module and the voltage at which it operates, illustrated in Fig 2.4
latterly (as used in this book)	Applies to the period from 2000 onward; that is, after "the early PV era"
levelized cost of energy (LCOE)	The total cost of the generator over its lifetime divided by the total energy it delivers. This is expressed as $$\frac{\sum_{y=1}^{y=n} \left(F_y + M_y\right)/(1+r)^y}{\sum_{y=1}^{y=n} E_y/(1+r)^y}$$ where n is the lifetime in years, F is the cost of capital and financing, M is the cost of operation and maintenance, E is the energy produced, and r is the discount rate
Lomé Convention	A trade and aid agreement between the EC and various African, Caribbean, and Pacific (ACP) countries, first signed in February 1975 in Lomé, Togo
marginal cost of electricity	The costs experienced by system operator for the last kWh of electricity bought to match supply with demand
maximum power point tracker (MPPT)	A device to maintain the operating point of a solar array near to its maximum or peak power point, usually incorporated into the inverter or DC charge regulator; *not to be confused with* mechanical trackers, like single- and double-axis trackers, used to follow the sun
microcrystalline	Materials with a crystalline structure, where the size of each crystal is small – on the order of micrometers
micromorph or micromorphous	Hybrid solar cell materials incorporating layers of both microcrystalline and amorphous materials
monocrystalline	One crystal – monocrystalline structures are composed of a single crystal throughout
net metering	Arrangement that prices electricity exported to the grid at the same level as that for energy purchased

	from the grid by the user at the connection point (see Section 3.5)
of the day (of prices, costs, and financial value)	Not adjusted for inflation, simply expressed at the level that applied at the time, *as opposed to* "real terms"
oil crisis *or* oil shock	The first oil crisis of 1973 and second oil crisis of 1979–1980, as described briefly in Sections 1.1 and 1.2
open-circuit voltage (V_{OC})	The voltage across the output connections of a solar cell or module when not connected to an external circuit (i.e., delivering no current)
our time frame	The period between 1973 and 1999, being the primary focus for this book. Also referred to as *"the early PV era"* or *"the first solar generation"*; further the words *"latterly"* and *"subsequently"* are frequently applied to developments after our time frame
peak oil	The point when the consumption of oil overtakes the rate at which new oil reserves are discovered. From then on the available reserves are in decline
peak power point	The point on a solar cell or module's *I–V* characteristic where it delivers the maximum power, as described in Section 2.3
performance ratio	The ratio of actual output of a solar generating system to the rated peak output of its solar modules under prevailing solar radiation conditions
photovoltaics	The conversion of light into electricity based on the photovoltaic effect, as further described in Section 2.3
polycrystalline	Materials that are crystalline in nature, but comprise more than one crystal within the overall structure
power	The instantaneous output of a system measured in kilowatts or equivalent units
power tower	Colloquialism for one type of concentrated solar power
priority access (to the grid)	Priority sometimes given to certain generating plants for connection to the grid and for priority

despatch. *Sometimes simply used for* "priority despatch"

priority despatch

Priority sometimes given to certain (e.g., renewable) generating plants allowing them to export their output to the grid first (with others being constrained off at times when surplus power is available)

progress ratio

The cost reduction resulting from doubling of cumulative production volume. A progress ratio of 81% means that each time total volume (e.g., of solar cells) ever made doubles, costs decline to 81% of their previous level
(*Note:* This is therefore counterintuitive – the higher the ratio, the slower the progress. I would have called this a 19% ratio, if the cost improvement was 19%.)

PV module

A single unit incorporating a number of interconnected solar cells in a protective package, usually with a frame

real terms (of prices, costs, and financial value)

Inflation adjusted to express the values over a period of several years in comparative terms, in a specified currency at a given year

renewable energy

Energy flows, which are naturally replenished, that occur repeatedly in the environment and can be harnessed for human benefit

Renewable Portfolio Standards (RPS)

A renewable energy support mechanism used mainly in certain states of the United States and in Japan to encourage energy suppliers to produce a certain proportion of their energy from prescribed renewables

renewables

Renewable energy sources

retail electricity price

The price paid by electricity users for power purchased from the grid

retail grid parity

The point at which the cost of solar generation matches the retail electricity price

second oil crisis

The second oil shock brought about by the 1979 Iranian revolution, as outlined in Section 1.2

Shockley–Queisser efficiency limit

The maximum efficiency of a solar cell as calculated according to theoretical physics, initially

	by William Shockley and Hans Queisser (see Section 12.2)
short-circuit current (I_{SC})	The current delivered by a solar cell or module, when its output contacts are directly connected together (i.e., at zero voltage)
single-axis tracker	A tracker that follows the sun across the sky in its daily orbit (but does not adjust for the seasonal height of the sun)
solar cell	A semiconductor device that converts light into electricity
solar farm *or* solar park	A large-scale solar power installation for merchant power generation
solar module	See PV module
Solar Power book	*Solar Power for the World: What You Wanted to Know About Photovoltaics* [43], edited by Wolfgang Palz[qv]
stand-alone (power systems)	Configured to deliver energy to a local user or applications, as opposed to grid-connected systems
standard test conditions (STC)	for solar cell and module testing (see Section 12.3)
subarray	A number of electrically interconnected solar modules
subsequently (as used in this book)	Often applies to the period from 2000 onward, that is, after "our time frame"
terrestrial photovoltaics	The use of PV in the earth's atmosphere – as opposed to space applications
thin film (solar cells)	Solar cells comprising active layers deposited as films typically on the order of 1 μm thick. This includes cells of many different materials, including amorphous silicon, cadmium telluride, and copper–indium diselenide
tilt angle	The angle of a solar array from the horizontal
tracker	A device that allows the solar array to follow the sun (see separately maximum power point tracker)
transmission network	The electricity transmission system operating at high voltage
Trinity House	The agency responsible for the UK's navigational aids and lighthouses

variability	The noncontinuous nature of some forms of generation that depend on inputs that vary with climatic factors, such as sunlight levels, wind speed, or tidal flows
wholesale electricity price	The price paid to generators for electricity delivered into the grid
wholesale grid parity	The point at which the cost of solar generation matches the wholesale electricity price

Units, Abbreviations, and Conversion Factors

The following units are used in this book:

Area	Square meters (m^2) and hectares (ha)
Energy	Watt-hours (Wh); 1 Wh is a power of 1 W flowing for 1 h
Length	Meters (m) and related multiples
Money	US dollars ($) or relevant currency for local programs
Power	Watts (W) and related multiples
PV power	Watts peak (W_P – see Section 2.3) under STC (see Section 12.3)

These conversion factors show how the above units relate to each other and to other commonly used units, and the derivation of other ratios used in this book.

Power

1 µW; microwatt	10^{-6} watts; 1 millionth of a watt
1 mW; milliwatt	10^{-3} watts; 1 thousandth of a watt; 1000 µW
1 W; watt	= 1 newton meter (N m) per second
Watt (electrical)	W_e = volts × amps; for example, 1 amp at 1 volt = 1 watt
W_P (watts peak)	Output of a solar device at peak power point under STC
W_P (watts peak)	$W_P = V_{PP} \times I_{PP}$
1 kW; kilowatt	10^3 watts; 1000 W
1 kWe	$\cong$ 1.34 horsepower
1 MW; megawatt	10^6 watts; 1 million watts; 1000 kW
1 GW; megawatt	10^9 watts; 1000 MW
1 TW; terawatt	10^{12} watts; 1000 GW

Energy

1 J; joule	1 watt flowing for 1 second
1 Wh; watt-hour	1 watt flowing for 1 hour. 1 Wh = 3600 joules
1 kWh; kilowatt-hour	1 kilowatt flowing for 1 hour $\cong$ 3412 Btu (British thermal units)
1 MWh; megawatt-hour	1 megawatt flowing for 1 hour $\cong$ 0.086 tonnes of oil equivalent (Toe)
1 GWh; gigawatt-hour	1 gigawatt flowing for 1 hour $\cong$ 86 tonnes of oil equivalent (Toe)
1 Toe; tonne of oil equivalent	$\cong$ 11.86 MWh

Peak Sun Hours: Capacity Factors

1 peak sun hour per day	Is a capacity factor of $1/24 \cong 4.167\%$
1000 peak sun hours per year	Is a capacity factor of $1000/365/24 \cong 11.416\%$

Length

1 μm; micrometer	10^{-6} meters; 1 millionth of a meter; 1/1000 mm
1 mm; millimeter	10^{-3} meters; 1 thousandth of a meter; 1000 μm
1 cm; centimeter	10^{-2} meters; 1 hundredth of a meter; 10 mm
1 m; meter	$\cong$ 39.4 inches; $\cong$ 3.28 feet; $\cong$ 1.09 yards
1 km; kilometer	10^3 meters; 1000 m; also $\cong$ 0.62 miles

Area

1 ha; hectare	10^4 square meters; $\cong$ 100 m × 100 m square; $\cong$ 2.47 acres
1 km^2; square kilometer	100 hectares; also $\cong$ 0.386 square miles

Money

US$1	At the time of writing, approximately €0.8; ¥110; £0.7

E

Index of Detailed Table of Contents

F

Indexes of Figures, Images, and Tables

Many people have provided and given permission to use photographs and images, and these are acknowledged under the "Image credits" section, and cited in the relevant references.

Images listed without reference in the following list are from the author. Uncaptioned photographs in the text were kindly provided by the individual or organization shown [25], Intersolar Group [27], Bernard McNelis [68], or the author.

Index of Captioned Figures and Images

As already mentioned, Section 10.2 contains many uncaptioned photographs, which are not included in the index below.

The Solar Generation: Childhood and Adolescence of Terrestrial Photovoltaics, First Edition. Philip R. Wolfe.
© 2018 by the Institute of Electrical and Electronic Engineers, Inc. Published 2018 by John Wiley & Sons, Inc.

Index of Tables

Index

The Solar Generation: Childhood and Adolescence of Terrestrial Photovoltaics, First Edition. Philip R. Wolfe.
© 2018 by the Institute of Electrical and Electronic Engineers, Inc. Published 2018 by John Wiley & Sons, Inc.

.